T0320761

Optimal Control of Distributed Nuclear Reactors

MATHEMATICAL CONCEPTS AND METHODS
IN SCIENCE AND ENGINEERING

Series Editor: Angelo Miele
>*Mechanical Engineering and Mathematical Sciences*
>*Rice University*

Recent volumes in this series:

A Continuation Order Plan is available for this series. A continuation order will bring delivery of
each new volume immediately upon publication. Volumes are billed only upon actual shipment.
For further information please contact the publisher.

Optimal Control of Distributed Nuclear Reactors

G. S. Christensen
University of Alberta
Edmonton, Alberta, Canada

S. A. Soliman
Ain Shams University
Cairo, Egypt

and

R. Nieva
Instituto de Investigaciones Electricas
Cuernavaca, Morelos, Mexico

Plenum Press • New York and London

Library of Congress Cataloging-in-Publication Data

Christensen, G. S. (Gustav S.)
    Optimal control of distributed nuclear reactors / G.S.
Christensen, S.A. Soliman, and R. Nieva.
        p.   cm. -- (Mathematical concepts and methods in science and
engineering ; 41)
    Includes bibliographical references.
    ISBN 0-306-43305-2
    1. Nuclear reactors--Control--Mathematical models.   I. Soliman,
S. A.   II. Nieva, R.   III. Title.   IV. Series.
TK9202.C48   1990
621.48'35'011--dc20   89-29220
                                                              CIP

© 1990 Plenum Press, New York
A Division of Plenum Publishing Corporation
233 Spring Street, New York, N.Y. 10013

Printed in the United States of America

To our wives and parents

Penelope (G.S.C.)
Laila (S.A.S.)
Sara and David (R.N.)

Preface

This book is devoted to the mathematical optimization theory and modeling techniques that recently have been applied to the problem of controlling the shape and intensity of the power density distribution in the core of large nuclear reactors.

The book has been prepared with the following purposes in mind:

1. To provide, in a condensed manner, the background preparation on reactor kinetics required for a comprehensive description of the main problems encountered in designing spatial control systems for nuclear reactor cores.
2. To present the work that has already been done on this subject and provide the basic mathematical tools required for a full understanding of the different methods proposed in the literature.
3. To stimulate further work in this challenging area by weighting the advantages and disadvantages of the existing techniques and evaluating their effectiveness and applicability.

In addition to coverage of the standard topics on the subject of optimal control for distributed parameter systems, the book includes, at a mathematical level suitable for graduate students in engineering, discussions of concepts of functional analysis, the representation theory of groups, and integral equations.

Although these topics constitute a requisite for a full understanding of the new developments in the area of reactor modeling and control, they are seldom treated together in a single book and, when they are, their presentation is often directed to the mathematician. They are thus relatively unknown to the engineering community.

Our audience consists mainly of graduate engineers and applied scientists concerned with the application of mathematical techniques to distributed parameter systems.

The body of this book consists of Chapters 2–6. Chapter 2 provides background preparation and provides us with the necessary tools to define the optimization problems. In Section 2.2, we review matrix theory. In Section 2.3, we discuss the calculus of variations as an optimization technique applied to a distributed system, and we explain the isoperimetric problem. Section 2.4 introduces dynamic programming and the principle of optimality for continuous processes. In Section 2.5, we consider Pontryagin's maximum principle. Finally, Section 2.6 discusses some powerful minimum norm problems of functional analysis, some of which have been applied in this book.

Chapter 3 is concerned with model decomposition techniques and model expansion techniques and their application to distributed parameter systems. One decomposition approach is discussed, namely, decomposition according to different time constants. Recent results obtained by the authors are based on these principles. A decomposition technique based on model reduction by means of symmetry considerations is discussed. Here we review the basic rudiments of the representation theory of groups and symmetry principles. Some recent results obtained by the authors based on such principles are reported here.

Chapter 4 is devoted to the problem of controlling the neutron flux distribution in a nuclear reactor core, in which spatial kinetic effects are important. In Section 4.3, the optimization technique of the minimum norm problem in Hilbert spaces is applied to the problem of adjusting the neutron flux for a general distributed nuclear reactor whose dynamic behavior is described in the neighborhood of an equilibrium condition. In Section 4.4, the maximum principle is applied to a linear deterministic mathematical model for a nuclear reactor system. Section 4.5 discusses the application of singular perturbation theory to the optimal control of nuclear reactors to obtain a suboptimal control law with spatially distributed parameters. The method proposed in this section is based on the boundary layer method connected with a model expansion analysis utilizing the Helmholtz mode, in which the inverse of the neutron velocity is a small quantity. Finally, in Section 4.6, we discuss the application of the system tau method, which is used to solve systems of coupled ordinary differential equations with initial, boundary and/or intermediate conditions. Optimal control problems of nuclear systems can be solved by this technique and by using Pontryagin's maximum principle when the system state equations and their associated performance index are transferred into a system of coupled differential equations with mixed boundary conditions.

Chapter 5 addresses different techniques used to control nuclear reactors during load-following. Section 5.2 discusses the application of multistage mathematical programming, when the problem of controlling the total power and power distribution in a large reactor core to follow a known time-varying load schedule has been formulated as a multistage optimization problem. The control problem is solved subject to hard constraints, based on a three-dimensional linearized model with some slightly nonlinear features; the optimal control problem is solved by quadratic programming. Section 5.3 is devoted to the solution of the problem of control of xenon spatial oscillation in the load-following operation of a nuclear reactor. The problem is formulated as a liner-quadratic tracking problem in the context of modern optimal control theory, and the resulting two-point boundary problem is solved directly by the technique of initial value methods. In Section 5.4, we discuss another multilevel method used to control nuclear reactors having practical operational constraints and thermal limits. Due to the very large size of the problem, a decomposition is made using hierarchical control techniques. The optimization of the resulting subproblems is performed using the feasible direction method.

Chapter 6 is devoted to the applications of the minimum norm optimization techniques to the problem of controlling the shape and intensity of the power density distribution. The problem formulations considered here are all variations on the same theme: transform the state of the reactor close to the desired state distribution while minimizing a performance index that penalizes both the control effort and the deviations of the reactor state from a desired distribution. Constraints are imposed on the total power generated in the core and on the state of the reactor. These constraints relate to the situation in which it is desired to change or adjust the total power output from the core while minimizing the distortion of the neutron flux distribution. The problem of reducing xenon oscillations in minimum time is also discussed.

The final chapter summarizes the contents of the book and discusses the need for future research in this field.

G. S. Christensen
Edmonton, Alberta, Canada
S. A. Soliman
Cairo, Egypt
R. Nieva
Cuernavaca, Morelos, Mexico

Acknowledgments

The authors would like to acknowledge encouragement received in writing this book from Dr. P. R. Smy, Chairman, Department of Electrical Engineering, University of Alberta; Dr. M. E. El-Hawary, Professor, Department of Electrical Engineering, Nova Scotia Technical University; and Dr. A. K. El-Kharishy, Chairman, Electrical Power and Machines Department, Ain Shams University. We are grateful to our many friends and colleagues and, in particular, to Drs. D. H. Kelly, M. N. Oguztoreli, V. Gourishankar, P. Palting, F. O. Moeck, G. M. Frescura, M. S. Lasheen, A. M. El-Arabaty, and M. S. Morsy, Mr. P. D. McMorran, and Mr. J. C. Luxat. We wish to express our thanks to Ms. Barbara J. Peck, Director, Canterbury Executive Services, for her professional expertise in typing many drafts of the manuscript.

Contents

4. Optimal Control of Distributed Nuclear Reactors

1

Introduction

1.1. Nuclear Reactor Operation and Design

The thermal power distribution in the core of a large nuclear reactor depends on the fission-reaction rate at each point of the core; it is neither homogeneous nor stationary.

The rate at which fission reactions are produced is influenced by the neutron flux distribution. In turn, this distribution is affected by any change in the way neutrons are absorbed, diffused, or produced throughout the core. These changes may be induced by a variety of causes, among which are power-level adjustments, on-power refueling operations, fuel consumption, and accumulation of fission products in nuclear fuel.

The neutron flux distribution in most reactor designs is controlled by the introduction of (or extraction of) a neutron-absorbent substance at specific locations of the core. The main objective for the reactor control system is to maintain this neutron flux distribution within acceptable limits in order to avoid potentially harmful spots with high power density.

Optimal control theory can improve the performance of existing reactor control systems, and considerable research has been done in this area. This book is a contribution to that research.

1.2. Outline of the Book

The body of this book consists of Chapters 2–6. Chapter 2 provides background and tools to define optimization problems. Section 2.2 reviews matrix theory. Section 2.3 discusses the calculus of variations as an optimization technique applied to a distributed system, and explains the isoperimetric problem. Section 2.4 introduces dynamic programming and the principle of optimality for continuous processes. Section 2.5 considers the Pontryagin maximum principles. And Section 2.6 discusses some powerful minimum norm problems of functional analysis, some of which have been applied in this book.

In Chapter 3 we study model decomposition techniques and model expansion techniques and their applications to distributed parameter systems, and we discuss decomposition according to widely different time constants. Recent results obtained by us are based on these principles. We also discuss a decomposition technique based on model reduction by means of symmetry considerations, in which we review the rudiments of the representation theory of groups and symmetry principles. Some recent results obtained by us are based on the principles reported here.

Chapter 4 examines the problem of controlling the neutron flux distribution in a nuclear reactor core, in which spatial kinetic effects are important. In Section 4.3, we apply the optimization technique of the minimum norm problem in Hilbert spaces to the problem of adjusting the neutron flux for a general distributed nuclear reactor whose dynamic behavior is described in the neighborhood of an equilibrium condition. In Section 4.4, we apply the maximum principle to a linear deterministic mathematical model of a nuclear reactor system. Section 4.5 discusses the application of singular perturbation theory to optimal control of nuclear reactors to obtain a suboptimal control law with spatially distributed parameters. The proposed method is based on the boundary layer method combined with a model expansion analysis that uses the Helmholtz mode, in which the inverse of the neutron velocity is a small quantity. Finally, in Section 4.6, we discuss the application of the system tau method (STM), which is used to solve a system of coupled ordinary differential equations with initial, boundary, and/or intermediate conditions. The optimal control problem of nuclear systems can be solved by this technique and by Pontryagin's maximum principle, in which the system state equations and their associated performance index are translated into a system of coupled differential equations with mixed boundary conditions.

Chapter 5 addresses techniques for controlling nuclear reactors during load-following. Section 5.2 discusses the application of multistage mathematical programming; the problem of controlling the total power and power distribution in a large reactor core to follow a known time-varying load schedule has been formulated as a multistage optimization problem. The control problem is solved subject to hard constraints, based on a three-dimensional linearized model with some slightly nonlinear features, but the optimal control problem is solved by quadratic programming. In Section 5.3 we solve the problem of controlling xenon spatial oscillation in the load-following operation of nuclear reactors. The problem is formulated as a linear-quadratic tracking problem in the context of modern optimal control theory. The resulting two-point boundary problem is solved directly by initial value methods. In Section 5.4, we discuss another multilevel method used to control nuclear reactors having practical operational constraints and thermal limits. Due to the very large size of the problem, a decomposition is made using hierarchical control technique. The optimization of the resulting subproblems is performed using the feasible direction method.

Chapter 6 applies minimum norm optimization techniques to the problem of controlling the shape and intensity of the power density distribution. The problem formulations considered here are all variations on the same theme; transform the state of the reactor close to the desired state distribution while minimizing a performance index that penalizes both the control effort and the deviations of the reactor state from a desired distribution. Constraints are imposed on the total power generated in the core and on the state of the reactor. These constraints relate to the situation in which we want to change or adjust the total power output from the core while minimizing the distortion of the neutron flux distribution. The problem of reducing xenon oscillations in minimum time is also discussed.

Chapter 7 summarizes the book and discusses the need for future research in this field.

Most of the material in this book originated from work done by our past and current associates. A number of *Nuclear Science and Engineering* papers are used as primary sources and are cited where appropriate.

2

Some Mathematical
Optimization Techniques

2.1. Introduction (Refs. 2.1–2.9)

This chapter addresses certain mathematical optimization techniques that have been applied to distributed parameter models. We discuss variational calculus, dynamic programming, Pontryagin's maximum principle and the minimum norm problems of functional analysis. Section 2.2 examines some functional concepts from matrix analysis and discusses the variational calculus as an optimization technique and its application to a distributed system. In Section 2.3, we study dynamic programming and the principle of optimality, Section 2.4 summarizes Pontryagin's maximum principle. In Section 2.5, we deal with the functional analytic technique of formulating optimization problems in the minimum norm form. Section 2.6 briefly treats some basic concepts from functional analysis before stating powerful versions of minimum norm problems.

2.2. A Review of Matrix Theory (Ref. 2.2)

The only sensible way to handle multidimensional matters is to use vector-matrix notation. Otherwise the notation is far too cumbersome and will cause us to lose sight of what is taking place.

2.2.1. Vectors

A column of numbers

$$x = \begin{bmatrix} x_1 \\ x_2 \\ \vdots \\ x_n \end{bmatrix} \qquad \text{or} \qquad x = \mathrm{col}(x_1, x_2, \ldots, x_n)$$

is called a *column n-vector.* The elements x_i are the *components* of the vector; x_i may be real or complex and a function of one or more variables (time is an example). If $x_1(t), x_2(t), \ldots, x_n(t)$ are functions of time, we have the nth-dimensional vector

$$x(t) = \text{col}[x_1(t), x_2(t), \ldots, x_n(t)]$$

Vectors x and y are equal if their respective components are equal; that is, $x_i = y_i$, $i = 1, \ldots, n$. Addition of vectors x and y is defined by $x_i + y_i$, $i = 1, \ldots, n$; thus,

$$x + y = \text{col}(x_1 + y_1, \ldots, x_n + y_n)$$

In general, for vectors x, y and z,

1. $x + y = y + x$ (commutative law).
2. $(x + y) + z = x + (y + z)$ (associative law).

Multiplication of a vector x by a scalar c_1 is defined by the relation

$$c_1 x = x c_1 = \text{col}(c_1 x_1, c_1 x_2, \ldots, c_1 x_n)$$

2.2.2. Matrices

An array of elements

$$A = \begin{bmatrix} a_{11} & a_{12} & & & a_{1n} \\ a_{21} & a_{22} & & & a_{2n} \\ \vdots & \vdots & \vdots & a_{ij} & \vdots & \vdots \\ a_{m1} & a_{m2} & & & a_{mn} \end{bmatrix} \leftarrow \text{row } i$$

$$\uparrow$$
$$\text{column } j$$

is called a *matrix;* the numbers in the array are called the *elements* of A. The matrix A has m rows and n columns, and therefore is said to have *dimension* $m \times n$ or to be an $m \times n$ matrix. A useful shorthand for A is

$$A = [a_{ij}], \qquad i = 1, \ldots, m, \quad j = 1, \ldots, n$$

If $m = n$, the matrix A is called *square of order* n; otherwise A is *rectangular.* The elements of A may be real or complex numbers.

Two matrices A and B are equal if they have equal dimensions and all their elements in corresponding positions are identical; i.e., $a_{ij} = b_{ij}$ for all possible i and j.

2.2.2.1. Addition of Matrices

Two matrices A and B can be added only if they have the same dimensions and if their elements on the corresponding positions are added; for example, if $C = A + B$, then

$$c_{ij} = a_{ij} + b_{ij} \qquad \text{for all possible } i \text{ and } j.$$

It is clear that

1. $A + B = B + A$ (commutative law).
2. $A + (B + C) = (A + B) + C$ (associative law).

2.2.2.2. Multiplication by a Scalar

If α is a constant, then the product αA is obtained by multiplying every element of A by α:

$$\alpha[a_{ij}] = [\alpha a_{ij}]$$

2.2.2.3. Multiplication of Two Matrices

The product AB of matrices A and B is defined only if the number of columns of A is equal to the number of rows of B; BA and B are then said to be *conformable* for multiplication. The dimension of the resulting product can be found by the simple rule:

$$
\begin{array}{ccc}
A & B & = & C \\
(m \times n) & (n \times p) & & m \times p
\end{array}
$$

It is clear that a matrix is a symbol for a linear transformation. Hence, if $z = Ay$ and $y = Bx$ denote, respectively, the linear transformations

$$z_i = \sum_{j=1}^{N} a_{ij} y_j, \quad y_i = \sum_{j=1}^{N} b_{ij} x_j, \qquad i = 1, 2, \ldots, N$$

it is clear by direct substitution that the z_i are linear functions of the x_j because $z = A(Bx)$. Therefore it is natural to define AB as the matrix of this resultant linear transformation. So we define the ijth element of AB to be

$$\sum_{k=1}^{N} a_{i_k} b_{k_j}$$

Notice that, in general, $AB \neq BA$ even if BA is defined (i.e., matrix multiplication is not commutative). If $AB = BA$, then A and B are said to commute with each other.

The *identity matrix I* is a square matrix in which all the main-diagonal elements are 1's and all of the off-diagonal elements are 0's. In mathematical terms, this can be written as

$$a_{ii} = 1, \qquad a_{ij} = 0 \quad \text{for } i \neq j$$

For any $m \times n$ matrix A, it is easy to verify that

$$AI = A \qquad \text{and} \qquad IA = A$$

Matrix multiplication obeys the following general properties:

$$(AB)C = A(BC)$$
$$C(A + B) = CA + CB$$
$$(A + B)C = AC + BC$$
$$\alpha(AB) = (\alpha A)B = A(\alpha B)$$
$$(ABC)^T = C^T B^T A^T$$

2.2.2.4. Inverse Matrix

The inverse B of an $n \times n$ matrix A is defined to be a matrix satisfying

$$AB = BA = I$$

The usual notation for the inverse is A^{-1}. The classical method of computing A^{-1} involves determinants and cofactors of A. Three important results for nonsingular matrices are

(i) If A and B are nonsingular square matrices, then $(AB)^{-1} = B^{-1}A^{-1}$.

(ii) If A is nonsingular, then $AB = AC$ implies $B = C$.

(iii) If A is a nonsingular matrix with complex elements, then

$(A^*)^{-1} = (A^{-1})^*$, where * means the complex conjugate.

2.2.2.5. Partitioned Matrices (Ref. 2.5)

Partitioning is apppplied to large matrices so that matrix manipulations can be carried out on the smaller blocks. More importantly, when matrices are multiplied in partitioned form, the basic rule can be applied to the blocks as though they were single elements.

For example, the 3×5 matrix

$$A = \begin{bmatrix} -1 & 2 & 3 & 1 & 2 \\ 0 & 1 & 5 & 3 & 4 \\ \hline 3 & -1 & 2 & 1 & 0 \end{bmatrix} = \begin{bmatrix} B & C \\ D & E \end{bmatrix}$$

is partitioned into four blocks, where B, C, D, and E are the arrays indicated by the dashed lines. By convention, A can be regarded as a submatrix of itself.

If the partitioned matrix A is multiplied by the matrix X,

$$X = \begin{bmatrix} X_1 \\ X_2 \end{bmatrix}$$

then

$$AX = \begin{bmatrix} B & C \\ D & E \end{bmatrix} \begin{bmatrix} X_1 \\ X_2 \end{bmatrix} = \begin{bmatrix} BX_1 + CX_2 \\ DX_1 + EX_2 \end{bmatrix}$$

provided that the blocks are conformable for multiplication (i.e., the products BX_1, CX_2, etc., exist). Therefore in the product AX, the number of columns in each block of A must equal the number of rows in the corresponding block of X.

If A is square and its only nonzero elements can be partitioned as principal submatrices, then it is called *block diagonal*. For example, the matrix

$$A = \begin{bmatrix} 2 & 3 & 0 \\ 5 & -1 & 0 \\ \hline 0 & 0 & 3 \end{bmatrix}$$

is block diagonal. A convenient notation for this is to write A as

$$A = \text{diag}[A_1, A_2, \ldots, A_k]$$

where $A_1, A_2, \ldots, A_k$ are square matrices, not necessarily of equal dimension, on the main diagonal. The inverse A^{-1} of $A = \text{diag}(A_1, A_2, \ldots, A_k)$ is $A^{-1} = \text{diag}(A_1^{-1}, A_2^{-1}, \ldots, A_k^{-1})$.

2.2.2.6. Partitioned Matrix Inversion

It is difficult to obtain the inverse of a large matrix by the classical method. The partitioned form is much simpler to use. Let F be a partitioned matrix

$$F = \begin{bmatrix} A_{n \times n} & B_{n \times m} \\ C_{m \times n} & D_{m \times m} \end{bmatrix}$$

and let

$$F^{-1} = \begin{bmatrix} W_{n \times n} & X_{n \times m} \\ Y_{m \times n} & Z_{m \times m} \end{bmatrix}$$

By definition, $FF^{-1} = I$, so

$$\begin{bmatrix} A & B \\ C & D \end{bmatrix} \begin{bmatrix} W & X \\ Y & Z \end{bmatrix} = \begin{bmatrix} I_n & 0 \\ 0 & I_m \end{bmatrix}$$

Applying the rule of partitioned multiplication gives

$$AW + BY = I_n$$

$$AX + BX = 0$$

$$CW + DY = 0$$

$$CX + DZ = I_m$$

Solving these equations, we obtain

$$W = A^{-1} - A^{-1}By$$

$$Y = -(D - CA^{-1}B)^{-1}CA^{-1}$$

$$Z = (D - CA^{-1}B)^{-1}$$

$$X = -A^{-1}B(D - CA^{-1}B)^{-1}$$

provided that A is nonsingular.

2.3. Calculus of Variations (Ref. 2.6)

The calculus of variations is a powerful method for solving a variety of problems. We introduce variational calculus by deriving the Euler equations and their associated transversality conditions.

Consider a continuous differentiable function $x(t)$. We want to find a function $x(t)$ that minimizes the performance index

$$J[x(t)] = \int_{t_0}^{t_f} F[x(t), \dot{x}(t), t] \, dt \qquad (2.1)$$

We introduce an arbitrary continuous differentiable function $\eta(t)$ with the stipulation that if $x(t_0)$ is specified as part of the problem the $\eta(t_0)$ must vanish; similarly if $x(t_f)$ is specified, then $\eta(t_f)$ must vanish. If we then define a new function

$$x(t) = \bar{x}(t) + \varepsilon\eta(t) \qquad (2.2)$$

where $\bar{x}(t)$ is an optimum, the parameter ε is real positive. Now we can write the functional of equation (2.1) as

$$J(\varepsilon) = \int_{t_0}^{t_f} F[\bar{x}(t) + \varepsilon\eta(t), \dot{\bar{x}}(t) + \varepsilon\dot{\eta}(t), t] \, dt \qquad (2.3)$$

It follows that the function $J(\varepsilon)$ must take on its minimum at $\varepsilon = 0$, where its derivative must vanish:

$$\frac{dJ(\varepsilon)}{d\varepsilon}\bigg|_{\varepsilon=0} = 0 \tag{2.4}$$

From the calculus of variations we have

$$\frac{dJ(\varepsilon)}{d\varepsilon} = \frac{\partial J(\varepsilon)}{\partial x(t)}\frac{dx(t)}{d\varepsilon} + \frac{\partial J(\varepsilon)}{\partial \dot{x}(t)}\frac{d\dot{x}(t)}{d\varepsilon} + \frac{\partial J(\varepsilon)}{\partial t}\frac{dt}{d\varepsilon} \tag{2.5}$$

$$= \frac{\partial J(\varepsilon)}{\partial x(t)}\eta(t) + \frac{\partial J(\varepsilon)}{\partial \dot{x}(t)}\dot{\eta}(t) + 0 \tag{2.6}$$

We can differentiate under the integral sign in equation (2.3) to get

$$\frac{dJ(\varepsilon)}{d\varepsilon} = \int_{t_0}^{t_f} [F_x\eta(t) + F_{\dot{x}}\dot{\eta}(t)]\, dt \tag{2.7}$$

where

$$F_x = \frac{\partial J(x, \dot{x}, t)}{\partial x} \tag{2.8}$$

$$F_{\dot{x}} = \frac{\partial J(x, \dot{x}, t)}{\partial \dot{x}} \tag{2.9}$$

In equation (2.7), we integrate the second term by parts to obtain

$$\frac{dj(\varepsilon)}{d\varepsilon}\bigg|_{\varepsilon=0} = 0 = \int_{t_0}^{t_f}\left(F_x - \frac{d}{dt}F_{\dot{x}}\right)\eta(t)\, dt + [F_{\dot{x}}\eta(t)]_{t_0}^{t_f} \tag{2.10}$$

Equation (2.10) is zero only if the integrand vanishes identically, so we obtain

$$F_x - \frac{d}{dt}F_{\dot{x}} = 0, \qquad t_0 < t < t_f \tag{2.11}$$

$$[F_{\dot{x}}\eta(t)]_{t_f} - [F_{\dot{x}}\eta(t)]_{t_0} = 0 \tag{2.12}$$

Equation (2.11) is *Euler's differential equation*, and equation (2.12) is the *transversality condition*.

If $x(t_0)$ is specified, then, since $\eta(t_0)$ must be zero, $F_{\dot{x}}\eta(t)$ must vanish at $t = t_0$. If $x(t_f)$ is specified, then, since $\eta(t_f)$ must be zero, $F_{\dot{x}}\eta(t)$ must vanish at $t = t_f$, because at $x(t_0)$ and $x(t_f)$ we have $x(t_0) = \bar{x}(t_0)$ and $x(t_f) = \bar{x}(t_f)$. For the free-end-point problem, we have

$$F_{\dot{x}(t_f)} = 0, \qquad \eta(t_f) \neq 0 \tag{2.13}$$

2.3.1. The Isoperimetric Problem (Ref. 2.5)

Some optical control problems have constraint relationships between the scalar elements of the state trajectory, which occurs in many physical problems. The problem with such constraints can be formulated as follows.

Find the function $x(t)$ that maximizes the index

$$J = \int_{t_0}^{t_f} F[x(t), \dot{x}(t), t]\, dt \tag{2.14}$$

subject to an integral constraint of the form

$$\int_{t_0}^{t_f} g[x(t), \dot{x}(t), t] = b \tag{2.15}$$

We can form an augmented cost function by adjoining equation (2.15) to the cost functional in equation (2.14) via the Lagrange multiplier λ:

$$J = \int_{t_0}^{t_f} \tilde{F}[x(t), \dot{x}(t), t]\, dt \tag{2.16}$$

where

$$\tilde{F}(\cdot) = F(\cdot) + \lambda g(\cdot) \tag{2.17}$$

As a result, we obtain the modified Euler equation

$$\tilde{F}_x - \frac{d}{dt}(\tilde{F}_{\dot{x}}) = 0 \tag{2.18}$$

and the transversality condition

$$[\tilde{F}_{\dot{x}}\eta(t)]_{t_f} - [\tilde{F}_{\dot{x}}\eta(t)]_{t_0} = 0 \tag{2.19}$$

2.3.2. A Distributed System (Ref. 2.1)

In all the cases we have considered thus far, the Euler equations have been ordinary differential equations. Extending the methods of this section, which are based on differential calculus, to the study of distributed systems is straightforward. Suppose that x is a function of two independent variables, which we call t and z, and that $x(t, z)$ is completely specified when $t = t_0$ or t_f for all z and when $z = z_0$ or z_f for all t. The problem is to find $x(t, z)$ that minimizes the double integral

$$J[x(t, z)] = \int_{z_0}^{z_f} \int_{t_0}^{t_f} \left[\frac{1}{2}\left(\frac{\partial x}{\partial t}\right)^2 + \frac{1}{2}\left(\frac{\partial x}{\partial z}\right)^2 + \phi(x)\right] dt\, dz \tag{2.20}$$

where $\phi(x)$ is any once-differentiable function of x.

If we call $\bar{x}(t, z)$ the optimum, we may write

$$x(t, z) = \bar{x}(t, z) + \varepsilon\eta(t, z) \tag{2.21}$$

where ε is a small number and $\eta(t, z)$ is a function that vanishes at $t = t_0$ or t_f and $z = z_0$ or z_f. For a particular function η the integral in equation (2.20) depends only on ε and may be written

$$J[x(t, z)] = \int_{z_0}^{z_f} \int_{t_0}^{t_f} \left[\frac{1}{2}\left(\frac{\partial \bar{x}}{\partial t} + \varepsilon \frac{\partial \eta}{\partial t}\right)^2 + \frac{1}{2}\left(\frac{\partial \bar{x}}{\partial z} + \varepsilon \frac{\partial \eta}{\partial z}\right)^2 + \phi(\bar{x} + \varepsilon \eta) \right] dt\, dz$$

(2.22)

The minimum of J occurs when $\varepsilon = 0$, by the definition of $\bar{x}$, and at $\varepsilon = 0$ the derivative of J with respect to ε must vanish. Thus,

$$\frac{dJ}{d\varepsilon}\bigg|_{\varepsilon = 0} = \int_{z_0}^{z_f} \int_{t_0}^{t_f} \left[\frac{\partial \bar{x}}{\partial t} \cdot \frac{\partial \eta}{\partial t} + \frac{\partial \bar{x}}{\partial z} + \dot{\phi}(\bar{x})\eta \right] dt\, dz = 0$$

(2.23)

But

$$\int_{t_0}^{t_f} \frac{\partial \bar{x}}{\partial t} \cdot \frac{\partial \eta}{\partial t} \, dt = \frac{\partial \bar{x}}{\partial t} \eta \bigg|_{t_0}^{t_f} - \int_{t_0}^{t_f} \frac{\partial^2 \bar{x}}{\partial t^2} \eta \, dt$$

(2.24)

and

$$\int_{z_0}^{z_f} \frac{\partial \bar{x}}{\partial z} \cdot \frac{\partial \eta}{\partial z} \, dz = \frac{\partial \bar{x}}{\partial z} \eta \bigg|_{z_0}^{z_f} - \int_{z_0}^{z_f} \frac{\partial^2 \bar{x}}{\partial z^2} \eta \, dz$$

(2.25)

The first terms in equations (2.24) and (2.25) vanish because of the restrictions on η. Now equation (2.23) becomes

$$\int_{z_0}^{z_f} \int_{t_0}^{t_f} \left[-\frac{\partial^2 \bar{x}}{\partial t^2} - \frac{\partial^2 \bar{x}}{\partial z^2} + \dot{\phi}(\bar{x}) \right] \eta(t, z) \, dt\, dz = 0$$

(2.26)

The integrand in equation (2.26) is equal to zero if and only if

$$\frac{\partial^2 x}{\partial t^2} + \frac{\partial^2 x}{\partial z^2} - \dot{\phi}(x) = 0$$

(2.27)

Equation (2.27) is the Euler partial differential equation for a distributed system.

2.4. Dynamic Programming and the Principle of Optimality (Ref. 2.1)

In the previous section, we discussed how to apply the calculus of variations to find the optimal decision. Over the past four decades an alternative approach, known as *dynamic programming*, has been developed. The foundation of this technique is based on the principle of optimality, as formulated by Bellman:

An optimal policy has the property that whatever the initial state and initial decision are, the remaining decisions must constitute an optimal policy with regard to the state resulting from the first decision.

For illustration, assume that we have a system with N stages and the decision $u_1, u_2, \ldots, u_N$ is to be made to minimize a certain function $J(x^N)$ according to a certain input–output relation at each stage, say

$$x^n = f^n(x^{n-1}, u^n), \qquad n = 1, 2, \ldots, N \qquad (2.28)$$

Having chosen u^1 and thus determined x^1, we must choose the remaining decisions $u_2, u_3, \ldots, u_N$ so that $J(x_N)$ is maximized for that x_1. Similarly, having chosen $u_1, u_2, \ldots, u_{N-1}$ and thus determined x_{N-1}, we must choose the remaining decision u^N so that $J(x_N)$ is minimized for that x_{N-1}. If the choice u_1 happened to be the optimal first choice and the remaining decisions were not optimal with respect to that x_1, we could always make $J(x_N)$ smaller by choosing a new set of remaining decisions. A computational algorithm has been developed as a result of the principle of optimality. Suppose that we have determined x_{N-1}; the choice of the remaining decision simply involves searching over all values of u_N to minimize $J(x_N)$ or, substituting equation (2.28),

$$\min_{u_N} J[f^N(x_{N-1}, u_N)] \qquad (2.29)$$

Since we do not know what the proper value of x_{N-1} is, we can than tabulate for each x_{N-1} the minimizing value of u_N and the corresponding minimum value of J.

We now move back one stage and suppose that we have available x_{N-2}, for which we must find decisions u_{N-1} and u_N that minimize $J(x_N)$. A specification of u_{N-1} will determine x_{N-1}, and for any given x_{N-1} we already have tabulated the optimal u_N and the value of J. Thus, we need simply search over u_{N-1} to find the tabulated x_{N-1} that gives the minimum value of J.

We now repeat the process for x_{N-3}, using the table to choose u_{N-2} for x_{N-2}, etc., until we finally reach x_0. Since x_0 is known, we can then use the table to choose u_1 for x_1. This gives the optimal value for x_1; we then find u_2 from the table for x_2, etc. In this way the optimal sequence $u_1, u_2, \ldots, u_N$ is constructed for a given x_0 by means of a sequence of minimization over a single variable.

For analytical purposes it is helpful to develop the mathematical formalism which describes the tabulation procedure. At the last stage the minimum value of $J(x^N)$ depends only upon the value of x^{N-1}. Thus, we define a function S^N of x^{N-1} as

$$S^N(x^{N-1}) = \min_{u^N} J[f^N(x^{N-1}u^N)] \qquad (2.30)$$

Similarly with two stages to go, we can define

$$S^{N-1}(x^{N-2}) = \min_{u^{N-1}} \min_{u^N} J(x^N) \qquad (2.31)$$

or, using equation (2.30),

$$S^{N-1}(x^{N-2}) = \min_{u^{N-1}} S^N[f^{N-1}(x^{N-2}, u^{N-1})] \qquad (2.32)$$

In general, then, we obtain the recursive relation

$$S^n(x^{n-1}) = \min_{u^n} S^{n+1}[f^n(x^{n-1}, u^n)] \qquad (2.33)$$

Consistent with this definition, we can define

$$S^{N+1}(x^N) = J(x^N) \qquad (2.34)$$

Equation (2.33) is a difference equation for the function S^n, with boundary condition given by equation (2.34), and is called the *Hamilton-Jacobi-Bellman equation*.

2.4.1. Dynamic Programming for Continuous Processes (Ref. 2.1)

In this section we discuss the application of dynamic programming to continuous processes. Consider the problem of finding the control vector $u(t)$ that minimizes

$$J = \int_0^{T_f} F[x(t), u(t)] \, dt \qquad (2.35)$$

and satisfying the system dynamics

$$\dot{x}(t) = f[x(t), u(t)] \qquad (2.36)$$

and $x(0) = b$.

If $g(b, t_f)$ is the minimum value of J, then we have

$$g(b, t_f) = \min_{u[0,t_f]} \int_0^{t_f} F[x(t), u(t)] \, dt \qquad (2.37)$$

This integration can be divided into two integrals:

$$g(b, t_f) = \min_{u[0,\Delta]} \min_{u[\Delta,t_f]} \left(\int_0^\Delta F\,dt + \int_\Delta^{t_f} F\,dt \right)$$

or

$$= \min_{u[0,\Delta]} \left(\int_0^\Delta F\,dt + \min_{u[\Delta,t_f]} \int_\Delta^{t_f} F\,dt \right) \tag{2.38}$$

If Δ is a small value of time, then, by using equation (2.36), we can write

$$\tilde{b} = b + \int_0^\Delta f\,dt \tag{2.39}$$

Using equation (2.37), we can write the second term in equation (2.38) as

$$g(\tilde{b}, \tilde{t}_f) = \min_{u[\Delta,t_f]} \int_\Delta^{t_f} F\cdot dt \tag{2.40}$$

where

$$t_f = \tilde{t}_f - \Delta \tag{2.41}$$

Substituting equation (2.40) in equation (2.38), we have

$$g(b, t_f) = \min_{u[0,\Delta]} \left[\int_0^\Delta F\,dt + g(\tilde{b}, \tilde{t}_f) \right] \tag{2.42}$$

Since Δ is a small value of time, we can use

$$\int_0^\Delta F\,dt = F[b, u(0)]\Delta \tag{2.43}$$

The function $g(b, t_f)$ can be expanded by a Taylor expansion to give

$$g(\tilde{b}, \tilde{t}_f) = g(b, t_f) + \Delta b \left[\frac{\partial g(b, t_f)}{\partial b} \right] - \Delta \left[\frac{\partial g(b, t_f)}{\partial T_f} \right] \tag{2.44}$$

Inserting equations (2.43) and (2.44) in equation (2.42), we have

$$g(b, t_f) = \min_{u[0,\Delta]} \left\{ F[b, u(0)]\Delta + g(b, t_f) + f[b, u(0)]\Delta \right.$$
$$\left. \cdot \left[\frac{\partial g(b, t_f)}{\partial b} \right] - \Delta \left[\frac{\partial g(b, t_f)}{\partial t_f} \right] \right\} \tag{2.45}$$

This equation can be written as

$$0 = \min_{u[0,\Delta]} \left\{ F[b, u(0)]\Delta + f[b, u(0)]\Delta \left[\frac{\partial g(b, t_f)}{\partial b} \right] - \Delta \left[\frac{\partial g(b, t_f)}{\partial t_f} \right] \right\} \tag{2.46}$$

because $g(b, t_f)$ is independent of the choice of u. In the limit as $\Delta \to 0$, we obtain the functional equation

$$\frac{\partial g(b, t_f)}{\partial t_f} = \min_y \left\{ F(b, y) + f(b, y) \left[\frac{\partial g(b, t_f)}{\partial b} \right] \right\} \qquad (2.47)$$

in which $y = u(0)$. The initial condition is

$$g(b, 0) = 0 \qquad (2.48)$$

Note that we can obtain the minimum in equation (2.47) by using calculus and search techniques, which may avoid difficulties in handling inequality constraints and in asserting that a global optimum is reached. The main drawback of the method is the large memory requirements in large-scale systems, which Bellman called the "curse of dimensionality."

2.5. Pontryagin's Maximum Principle (Refs. 2.2, 2.6, and 2.9)

Let a system be described by the dynamic equation

$$\dot{x}(t) = f[x(t), u(t), t] \qquad (2.49)$$

where $u(t)$ is an admissible control and $x(t)$ is the corresponding trajectory of the system. Let $x(t_0) = x(0)$, t_0 and t_f be specified, and $x(t_f)$ be free. The necessary conditions for $u(t)$ to be an optimum while minimizing the functional

$$J = G[x(t_f, t_f)] + \int_{t_0}^{t_f} L\{x(t), u(t), t]\} \, dt \qquad (2.50)$$

are the following two.

(1) There exists a function or vector $\lambda(t)$ such that $x(t)$ and $\lambda(t)$ are the solution of the equations

$$\dot{x}(t) = \frac{\partial H}{\partial \lambda(t)} \qquad (2.51)$$

$$\dot{\lambda}(t) = -\frac{\partial H}{\partial x(T)} \qquad (2.52)$$

subject to the boundary conditions

$$x(t_0) = x(0) \qquad (2.53)$$

$$\lambda(t_f) = \frac{\partial G(\cdot)}{\partial x(t)} \bigg|_{t=t_f} \qquad \text{at } x(t) = x(t_f) \qquad (2.54)$$

Elements of the Lagrange multiplier vector $\lambda(t)$ are often called *costates* or *adjoint variables*. The scalar function H is called the *Hamiltonian* and is given by

$$H[x(t), u(t), \lambda(t), t] = L[x(t), u(t), t] + \lambda^T(T)f[x(t), u(t), t] \quad (2.55)$$

(2) If the admissible control vector $u(t)$ is unrestricted, the functional $H[x(t), u(t), \lambda(t), t]$ has a local minimum at

$$\frac{\partial H}{\partial u(t)} = 0 \quad\quad\quad (2.56)$$

Most practical problems have inequality constraints on the control vector $u(t)$ and the states $x(t)$; in this case we are not free to apply equation (2.56). The maximum principle addresses this difficulty. In place of equation (2.56), the necessary condition is that the Hamiltonian function $H[x(t), u(t), \lambda(t), t]$ has an absolute minimum as a function of $u(t)$ over the admissible region Ω for all t in the interval (t_0, t_f). This condition can be expressed as the inequality

$$H[x(t), u^*(t), \lambda(t), t] \leq H[x(t), u(t), \lambda(t), t] \quad\quad (2.57)$$

where $u^*(t)$ is the optimal control vector in Ω.

2.6. Minimum Norm Problems of Functional Analysis (Refs. 2.5, 2.7, and 2.8)

In contrast to other optimization techniques, which generally approach problems of a particular dynamic nature, functional analysis with its geometric character provides a unified framework for discrete, continuous, distributed, or composite optimization problems.

The aim of this section is to discuss and review a class of functional-analytical problems that form part of the well-developed optimization theory by vector-space methods, namely, abstract minimum norm problems. First, however, we briefly discuss relevant concepts from functional analysis.

2.6.1. Some Concepts of Functional Analysis (Ref. 2.5)

The concepts presented are linked to portions of subsequent chapters.

2.6.1.1. Norms and Inner Products

A *norm* $\|\cdot\|$ is a real positive-definite value satisfying the following axioms:

1. $\|x\| \geq 0$ for all $x \in X$, $\|x\| = 0 \leftrightarrow x = 0$.

2. $\|x + y\| \le \|x\| + \|y\|$ for each $x, y \in X$.
3. $\|x\| = \alpha \|x\|$ for all scalars α and each $x \in X$.

A *normed linear space* X is a linear space in which every vector x has a norm (length). The norm functional is used to define a distance and a convergence measure

$$d(x, y) = \|x - y\|$$

For example, the space of continuous functions $x(t)$ on $[0, t]$, where $[0, T]$ is a closed bounded interval, can have one of the following norms:

$$\|x\|_1 = \int_0^T |x(t)| \, dt$$

$$\|x\|_2 = \left(\int_0^T |x(t)|^2 \, dt \right)^{1/2}$$

Let X be a linear space. A rule that assigns a scalar $\langle x, y \rangle$ to every pair of elements $x, y \in X$ is an *inner product* if the following conditions are satisfied:

1. $\langle x, y \rangle = \langle y, x \rangle$.
2. $\langle \alpha x + \beta y, z \rangle = \alpha \langle x, z \rangle + \beta \langle y, z \rangle$.
3. $\langle \lambda x, y \rangle = \lambda \langle x, y \rangle$.
4. $\langle x, x \rangle \ge 0$, $\langle x, x \rangle = 0 \leftrightarrow x = 0$.
5. $\langle x, x \rangle = \|x\|^2$.

The inner product in condition 1 can be written in component form as

$$\langle x, y \rangle = \sum_{i=1}^N x_i y_i$$

If $\langle x, y \rangle = 0$, we say that the vectors x and y are *orthogonal*. Furthermore, if A is a square matrix, then

$$\langle Ax, y \rangle = \sum_{i,j=1}^N a_{ij} x_i y_j$$

and thus

$$\langle Ax, y \rangle = \langle x, A^T y \rangle$$

A linear space X is called a *Hilbert space* if X is an inner product space that is complete with respect to the norm induced by the inner product. Equivalently, a Hilbert space is a *Banach space* whose norm is induced by an inner product. We now consider some examples of Hilbert spaces. The space $L_2[0, T]$ is a Hilbert space with inner product

$$\langle x, y \rangle = \int_0^T x(t) y(t) \, dt$$

Another example often occurs in this book. If $B(t)$ is a positive-definite matrix whose elements are functions of time, we can define the Hilbert space $L_{2B}^n(0, T_f)$. The inner product in this space is

$$\langle V(t), U(t) \rangle = \int_0^{T_f} V^T(t) B(t) U(t) \, dt$$

for every $V(t)$ and $U(t)$ in the space.

2.6.1.2. Transformations (Ref. 2.8)

A *transformation* is defined by a rule that associates with every element $x \in D$ an element $y \in Y$, where X and Y are linear vector spaces and $D \subset X$. If y corresponds to x under T, we write

$$y = T(x)$$

The transformation $T: X \to Y$ is *linear* if

$$T(\alpha_1 x_1 + \alpha_2 x_2) = \alpha_1 T(x_1) + \alpha_2 T(x_2)$$

for all $\alpha_1, \alpha_2 \in R$ (the real line) and for every $x_1, x_2 \in X$.

Let X and Y be normed spaces and let $T \in B(X, Y)$. The *adjoint* (*conjugate*) operation $T^*: Y^* \to X^*$ is defined by

$$\langle x, T^* y \rangle = \langle Tx, y^* \rangle$$

An important special case is that of a linear operator $T: H \to G$, where H and G are Hilbert spaces. If G and H are real, then they are their own duals and the operator T^* can be regarded as mapping G into H. In this case the adjoint relation becomes

$$\langle Tx, y \rangle = \langle x, T^* y \rangle$$

Note that the left-hand-side inner product is taken in G, while the right-hand-side inner product is taken in H.

A *composite transformation* can be formed as follows. Let $T: X \to Y$ and $G: Y \to Z$ be transformations. We define the transformation $GT: X \to Z$ by

$$(GT)(x) = G[T(x)]$$

We then say that GT is a composite of G and T, respectively.

2.6.2. Minimum Norm Problems

Historically, the method of moments is the first functional-analytical technique used in connection with optimal control problems, in which the minimum norm formulation is employed. The origin of the method of moments can be traced to very early developments in functional analysis. The problem of moments is stated below.

Problem 1. The *l*-Problem of Moments. Given the linearly independent elements f_i, $i = 1, 2, \ldots, N$, in a normed linear space B, the real numbers α_i, $i = 1, 2, \ldots, N$, and the positive real number l, find necessary and sufficient conditions for there to exist a linear functional u defined on B and satisfying the constraint

$$u(f_i) = \alpha_i, \qquad i = 1, 2, \ldots, N \tag{2.58}$$

with the norm

$$\|u\| = \sup_h \frac{|u(h)|}{\|h\|} \leq l \tag{2.59}$$

For comparison purposes, the *l*-problem of moments may be formulated in a different manner.

Problem 1(a). Let x be a given vector of the n-dimensional Euclidean space E_n. Let c be a convex set of linear functionals defined on the normed space B:

$$c = \left\{ u: \|u\| = \sup_h \frac{|u(h)|}{\|h\|} \leq l, h \in B \right\} \tag{2.60}$$

For a given linear bounded transformation F mapping c into E_n, find the necessary and sufficient conditions for there to exist a $u \in c$ satisfying

$$Fu = x \tag{2.61}$$

Problem 2. Let F be a compact linear bounded transformation mapping a Hilbert space H_1 into another Hilbert space H_2. For some x in H_2, minimize

$$\|Fu - x\|^2 \tag{2.62}$$

subject to u being in the sphere c in H_1 defined by

$$\|u\|^2 \leq l^2 \tag{2.63}$$

This transformation may be seen to be a variation of the *l*-problem of moments in the sense that the control u is not required to be a solution to equation (2.61), but only the best approximation in the H_2-norm.

The compactness assumption on F gave this problem a very general scope, since the range of R was no longer required to lie in a finite-dimensional space.

Solution to Problem 2. Either

$$\sup_{k>0} \|[F^*F + kI]^{-1}F^*x\| \leq l \tag{2.64}$$

in which case the sequence

$$u_k = [F^*F + kI]^{-1}F^*x \qquad (2.65)$$

is such that u_k converges to the optimal element u_0 of minimal norm

$$\lim_{k \to 0} \|Fu_k - x\|^2 = \inf_{u \in c} \|Fu - x\|^2 = \|Fu_0 - x\| \qquad (2.66)$$

or

$$\sup_{k \to 0} \|[F^*F + kI]^{-1}F^*x\| > l \qquad (2.67)$$

in which case

$$u_0 = [F^*F + k_0 I]^{-1}F^*x \qquad (2.68)$$

where k_0 is adjusted so that $\|u_0\| = l$ yields the unique solution to Problem 2.

The foregoing theory was extended to a Banach-space setting and discussed the questions of existence and uniqueness and the properties of optimal control. This extension can be formulated in the following minimum norm problem.

Problem 3. Let B and D be Banach spaces and I a bounded linear transformation defined on B with values in D. For each ξ in the range of T, find an element $u_\xi \in B$ that satisfies

$$\xi = Tu \qquad (2.69)$$

while minimizing the performance index

$$J(u) = \|u - \hat{u}\| \qquad (2.70)$$

Solution to Problem 3. The solution to this problem is

$$u_\xi = T^\dagger[\xi - t\hat{u}] + \hat{u} \qquad (2.71)$$

where the pseudoinverse operator for Hilbert spaces is

$$T^\dagger \xi = T^*[TT^*]^{-1}\xi \qquad (2.72)$$

provided that the inverse of TT^* exists.

The theorem as stated is an extension of the fundamental minimum norm problem where the objective functional is

$$J(u) = \|u\| \qquad (2.73)$$

The optimal solution for this case is

$$u_\xi = T^\dagger \xi \tag{2.74}$$

with $T^\dagger$ being the pseudoinverse associated with T.

The formulation of the previous problem may be seen to be a generalization of the l-problem of moments in that the range of T is not required to lie in a finite-dimensional space but in a general Banach space. The control u is required, however, as is the problem of moments, to satisfy a given equation.

We can use a Lagrange multiplier argument to obtain equation (2.71) for the optimal control vector u_ξ. We can obtain the augmented cost functional by adjoining to the cost functional in equation (2.70) the equality constraints in equation (2.69) via Lagrange's multiplier as follows;

$$J(u) = \|u - \hat{u}\|^2 + \langle \lambda, \xi - Tu \rangle, \qquad \lambda \in D \tag{2.75}$$

where λ is a Lagrange multiplier to be determined so that the constraints in equation (2.69) are satisfied. By utilizing properties of inner products, we can write

$$J(u) = \|u - \hat{u} - T^*(\lambda/2)\|^2 - \|T^*(\lambda/2)\|^2 + \langle \lambda, \xi \rangle \tag{2.76}$$

Only the first norm of equation (2.76) depends explicitly on the control vector u. To minimize J, we consider only

$$J(u) = \|u - \hat{u} - T^*(\lambda/2)\| \tag{2.77}$$

the minimum of J is achieved when

$$u_\xi = \hat{u} + T^*(\lambda/2) \tag{2.78}$$

To find the value of $\lambda/2$, we use the equality constraints

$$\xi = Tu_\xi \tag{2.79}$$

which gives

$$\lambda/2 = [TT^*]^{-1}[\xi - T\hat{u}] \tag{2.80}$$

Clearly, with an invertible TT^*, we write

$$u_\xi = T^*[TT^*]^{-1}[\xi - T\hat{u}] \tag{2.81}$$

which is the required result. If $u = 0$ in equation (2.81), we obtain

$$u_\xi = T^*[TT^*]^{-1}\xi \tag{2.82}$$

which is the same result obtained for the fundamental minimum norm problem.

Several variations on the general minimum energy problem which dramatically enlarged the application scope of the abstract minimum norm formulation were considered by Porter (Ref. 2.8). The most general of these variations is given in the next problem.

Problem 4. Let F be a bounded linear transformation from a Banach space B into a Banach space B_1, let T be a bounded linear transformation from B onto (or with dense range) a Banach space D, and let $\hat{u}$, $\hat{y}$, and ξ be given vectors in B, B_1, and D, respectively. Find u in B satisfying

$$Tu = \xi \tag{2.83}$$

that minimizes

$$\|u - \hat{u}\|^2 + \|Fu - \hat{y}\|^2 \tag{2.84}$$

Of particular interest to the application considered in this book is the Hilbert-space version of Problem 4, in which case the unique solution obtained by Porter and Williams is as follows.

Solution to Problem 4. The unique solution u_0 of the Hilbert space version of Problem 4 is

$$u_0 = [I + F^*F]^{-1}(T^+\eta + \hat{u} + F^*\hat{y}) \tag{2.85}$$

where η is the unique vector in D satisfying

$$\xi = T[I + F^*F]^{-1}(T^+\eta + \hat{u} + F^*\hat{y}) \tag{2.86}$$

F^* is the adjoint of F, and T^+ is the pseudoinverse of T defined by

$$T^+\xi = T^*(TT^*)^{-1}\xi \tag{2.87}$$

provided that TT^* is invertible.

References

2.1. BELLMAN, R., *Introduction to the Mathematical Theory of Control Process*, Vol. 1, Academic Press, New York, 1967.
2.2. BARNETT, S., *Matrix Methods for Engineers and Scientists*, McGraw-Hill, New York, 1979.
2.3. DENN, M. M., *Optimization by Variational Methods*, McGraw-Hill, New York, 1969.
2.4. DORNY, D. N., *A Vector Space Approach to Models and Optimization*, Wiley-Interscience, New York, 1975.
2.5. EL-HAWARY, M. E., and CHRISTENSEN, G. S., *Optimal Economic Operation of Electric Power Systems*, Academic Press, New York, 1979.
2.6. KIRK, D. E., *Optimal Control Theory: An Introduction*, Prentice-Hall, Englewood Cliffs, New Jersey, 1970.
2.7. LUENBERGER, D. G., *Optimization by Vector Space Methods*, Wiley, New York, 1969.
2.8. PORTER, W. A., *Modern Foundations of Systems Engineering*, MacMillan, New York, 1966.
2.9. SAGE, A., *Optimization System Controls*, Prentice-Hall, Englewood Cliffs, New Jersey, 1968.

3

Distributed Reactor Modeling

3.1. Introduction (Refs. 3.1–3.32)

There is a large difference in time between neutron kinetics and reactor poisoning dynamics; by assuming (validly) that neutron kinetics is an instantaneous process compared with the slow dynamics of the core poisoning effects, we can simplify the analysis of problems, such as xenon spatial stability or load-following, considerably. A very general reactor core model would then consist of the multigroup neutron diffusion equations at steady state coupled with the xenon and iodine dynamic equations.

We present a new method for obtaining an approximate solution to such a model. Our objective is to obtain an explicit functional relation between the neutron flux and the neutron absorption cross sections of the zone controllers. We want to reduce the number of variables involved in optimal control studies of load-following operations.

The proposed approach is described in terms of the one-energy group model with xenon and iodine dynamics. Several examples are presented in which we use data that correspond to a typical, pressurized-tube, heavy-water-modulated, large nuclear reactor.

The reactor models used in connection with the control of distributed nuclear reactors are usually classified as nodal, finite-difference, or modal expansion. This classification depends on which method is used to obtain an approximate description of the reactor core processes.

The nodal and modal expansion methods are the most widely employed for reactor control problems, in which use is made of distributed reactor models. The method of finite difference is used extensively and almost exclusively in reactor physics calculations.

Modal expansion has the following advantages over finite-difference and nodal methods:

(a) It can represent the distributed nature of the reactor core with as much detail as can the finite-difference approximation. However, far fewer equations are required.

(b) The number of equations required to represent the distributed reactor is comparable for the modal expansion and nodal methods. The spatial representation of the reactor core, is, however, more detailed for the modal expansion approach.

In general, the modal approach has two main disadvantages:

(a) For many modal expansion methods it is not possible to determine or even estimate error bounds between the true solution to the problem and the solution obtained by the modal expansion.

(b) For some modal expansion methods the spatial models are difficult to compute.

Several methods that use explicitly known, analytical functions as expansion modes have been proposed in the past to circumvent these disadvantages. The method of solutions functions of R. Bobone is probably the best example. The method was used successfully in critical calculations and also in computing the steady-state power distribution in a reactor core. Bobone's approach assumes the homogeneity of the diffusion coefficient and reactor buckling within each of several core regions. The method consists of expanding the neutron flux within each region in terms of solutions to the diffusion equation and then evaluating the expansion coefficients so as to minimize the mean squared error of the neutron flux and current at the region interfaces. Unfortunately, this method cannot be applied to the analysis of load-following operations, because the xenon distribution within the reactor regions makes the neutron absorption cross sections spatially and temporally dependent. Iwazumi and Koga proposed a method in which the spatially dependent parameters of the diffusion equation are replaced with constant terms that are chosen to keep the fundamental mode steady. The spatially dependent terms were then treated as the initial distribution of the control variable. The basic idea was to use the eigenfunctions of the Helmholtz operator as expansion modes. This approach was used by Owazumi and Koga in connection with the problem of changing the flux distribution from a given initial state to a desired final state in a short period of time while minimizing a functional that penalizes the terminal flux distribution error and the deviations of the control variable from the initial control distribution. The one-energy group model with one delayed neutron precursor in a slab reactor was considered in their work. Recently, Iwazumi's method was used by Asatani *et al.* in connection with the same control problem and the same reactor model.

Iwazumi's method has not been extended to treat load-following or xenon oscillation problems. Whether this extension would lead to justifiable simplifications is an open question.

3.2. The Multigroup Diffusion Equations

The method used to derive the multigroup diffusion equation is the modal expansion approximation in terms of analytical functions, and it is characterized by two important features:

1. The use of analytical functions as expansion modes makes it possible to bypass the problem of computing the expansion modes numerically.
2. The formalism of functional analysis, which is used in the development of the approximate model, makes it possible to estimate bounds on the approximating errors.

In deriving these equations, we consider three basic assumptions:

1. The geometrical configuration of the reactor core is such that a complete set of wave functions in $L_2[V]$ is explicitly known and has the property that the functions vanish at the extrapolated boundary of the core; $L_2[V]$ is the space of squared integrable functions over the core volume V.
2. The neutron diffusion coefficients for these different neutron energies are assumed to be homogeneous through the core. This assumption is acceptable for large reactor cores that, in addition to being lightly rodded, do not use reflectors or they have reflectors made of the same material that is used as the neutron moderator. The multigroup bucklings may be spatially dependent.
3. The reactor model is linearized in the neighborhood of a given power distribution, which could be at steady state or a slow transient during a large power-level change.

The description of the proposed approach is given in terms of the one-group model with xenon and iodine dynamics, which is presented by the following equations:

$$\nabla^2 \psi_\phi(\mathbf{r}, t) + b(\mathbf{r})\psi_\phi(\mathbf{r}, t) = f(\mathbf{r}, t), \qquad r \in V \tag{3.1}$$

$$b(\mathbf{r}) = \frac{1}{D}[v\Sigma_f(\mathbf{r}) - \Sigma_a(\mathbf{r}) - u_0(\mathbf{r}) - \sigma x_0(\mathbf{r})] \tag{3.2}$$

$$f(\mathbf{r}, t) = \frac{\sigma}{D}\phi_0(\mathbf{r})\psi_x(\mathbf{r}, t) + \hat{u}(\phi_0(\mathbf{r}), \mathbf{r}, t) \tag{3.3}$$

$$x_0(\mathbf{r}) = \frac{\gamma_I \Sigma_f(\mathbf{r})\phi_0(\mathbf{r})}{\lambda_x + \sigma\phi_0(\mathbf{r})} \tag{3.4}$$

$$\frac{\partial \psi_I}{\partial t}(\mathbf{r}, t) = -\lambda_I \psi_I(\mathbf{r}, t) + \gamma_I \Sigma_f(\mathbf{r}) \psi_\phi(\mathbf{r}, t) \tag{3.5}$$

$$\frac{\partial \psi_x}{\partial t}(\mathbf{r}, t) = \lambda_I \psi_I(\mathbf{r}, t) - [\lambda_x + \sigma\phi_0(\mathbf{r})]\psi_x(\mathbf{r}, t) - x_0(\mathbf{r})\sigma\psi_\phi(\mathbf{r}, t) \tag{3.6}$$

with initial conditions

$$\psi_I(\mathbf{r}, t_0) = \psi_{I0}(\mathbf{r}) \tag{3.7}$$

$$\psi_x(\mathbf{r}, t_0) = \psi_{x0}(\mathbf{r}) \tag{3.8}$$

and boundary condition at the core-extrapolated boundary (∂V)

$$\psi_\phi(\mathbf{r}, t) = 0 \tag{3.9}$$

Equations (3.1) to (3.9) describe the dynamic behavior of the neutron flux deviation ψ_ϕ and the deviation of xenon and iodine concentrations from their corresponding equilibrium distributions ϕ_0, X_0, and I_0. The parameters σ, λ_x, γ_I, and λ_I denote, respectively, the neutron absorption microscopic cross section of xenon-135, the decay constants of xenon and iodine, and the iodine yield per fission; $\hat{u}$ denotes the control function deviations from the equilibrium control u_0.

For clarity, the procedure used in deriving equations (3.1) to (3.9) has been divided into five steps.

1. The neutron diffusion equation, (3.1), is transformed into a non-homogeneous Fredholm integral operator equation of the second kind.
2. The kernel of the resulting integral operator is approximated by a degenerate kernel. An error bound on the $L_2[V]$ norm is obtained, the error between the true solution to the Fredholm equation and the solution obtained by means of the degenerate kernel approximation.
3. The integral equation is solved for the neutron flux by applying the method of degenerate kernels. An explicit relation for the neutron flux is obtained in terms of the xenon concentration and the control variables.
4. The explicit expression for the neutron flux is substituted in equations (3.5) and (3.6), and an approximate solution to these equations is obtained by expanding the xenon and iodine distributions in terms of a finite number of wave functions. A bound is found on the $L_2^2[V]$, and the Neumann series is used to represent the error.
5. Using the explicit expression for the neutron flux and the modal expansion solution to equations (3.5) and (3.6), we write the neutron flux in terms of control variables only.

3.2.1. Solution of the Diffusion Equation

The one-group neutron diffusion equation (3.1) is nonhomogeneous Helmholtz equation in the Hilbert space $L_2[V]$ with a space-dependent parameter $b(r)$, which, by assumption, is a piecewise continuous function of the spatial variable r. An approximate solution to equation (3.1) could be obtained by expanding the neutron flux in terms of a finite number of functions. This procedure is used extensively in the literature. The simplicity of this approach, however, has to be weighed against the disadvantage of the lack of error estimates. An alternative approach consists of transforming equation (3.1) into an equivalent integral equation that is amenable to both error analysis and computation.

The first step toward transforming (3.1) is to express $b(r)$ as the sum of two terms,

$$b(r) = b_0 + \hat{b}(r) \tag{3.10}$$

where b_0 is spatially independent. There are an infinite number of different ways to do this. The considerations that dictate the best choice will be discussed later. For the moment, we assume that b_0, in addition to being a real number, is also different from all the eigenvalues of the following boundary value problem in $L_2[V]$:

$$\nabla^2 \psi_j(r) = \lambda_j \psi_j(r), \qquad r \in V \tag{3.11}$$

and

$$\psi_j(r) = 0, \qquad r \in \partial V \tag{3.12}$$

where V is the region occupied by the reactor core, ∂V is the extrapolated boundary, and ∇^2 is a Laplacian operator. By assumption, the set of functions $\{\psi_j\}$ is explicitly known.

Using (3.10) in equation (3.1), we get

$$\nabla^2 \psi_\phi(r, t) + b_0 \psi_\phi(r, t) - \hat{b}(r) \psi_\phi(r, t) \tag{3.13}$$

Using the Green's function $G(r, r')$ corresponding to the left side of equation (3.13), we obtain

$$\psi_\phi(\mathbf{r}, t) = \int_V G(\mathbf{r}, r') f(r') \, dV' - \int_V G(\mathbf{r}, r') \hat{b}(\mathbf{r}') \psi_\phi(\mathbf{r}', t) \, dV' \tag{3.14}$$

where the integrals are over the core volume.

By invoking the first assumption of Section 3.2, we can show that the Green's function can be obtained in series form:

$$G(\mathbf{r}, r') = \sum_j \frac{\psi_j(\mathbf{r}) \psi_j(\mathbf{r}')}{b_0 + \lambda_j} \tag{3.15}$$

where $\{\lambda_j\}$ and $\{\psi_j\}$ form the eigensolution to equation (3.11).

Equation (3.14) is a Fredholm integral equation of the second kind in the function space $L_2[V]$. Also, since the integral operator in equation (3.14) is compact, it is clear that there exists a unique solution in $L_2[V]$ provided that -1 is not in the eigenvalue spectrum of the integral operator.

3.2.2. The Method of Degenerate Kernels

Of all the methods available for solving the nonhomogeneous Fredholm integral equation, the method of degenerate kernels is particularly attractive to use when the Green's function (3.15) can be adequately approximated by a finite series.

Substituting the Green's function (3.15) in the integral equation (3.14), we have

$$\psi_\phi(\mathbf{r}, t) = \sum_j \frac{\psi_j(\mathbf{r})}{b_0 + \lambda_j} \int_V \psi_j(\mathbf{r}')(f(r', t) - \hat{b}(\mathbf{r}')\psi_\phi(\mathbf{r}', t))\, dV' \qquad (3.16)$$

which can be rewritten as

$$\psi_\phi(\mathbf{r}, t) = \sum_j \frac{\psi_j(\mathbf{r})}{b_0 + \lambda_j}[e_j(t) - C_j(t)] \qquad (3.17)$$

where the coefficients $e_j(t)$ and $C_j(t)$, defined as

$$C_j(t) = \int_V \psi_j(\mathbf{r})\hat{b}(\mathbf{r})\psi_\phi(\mathbf{r}, t)\, dV \qquad (3.18)$$

$$e_j(t) = \int_C \psi_j(\mathbf{r})f(\mathbf{r}, t)\, dV \qquad (3.19)$$

satisfy the algebraic system of equations

$$C_i(t) = \sum_j d_{ij}(e_j - C_j) \qquad (3.20)$$

where

$$d_{ij} = \int_C \frac{\psi_i(\mathbf{r})\hat{b}(\mathbf{r})\psi_j(\mathbf{r})\, dV}{b_0 + \lambda_j} \qquad (3.21)$$

A natural approach for solving equation (3.16) consists of approximating the Green's function by a finite series and then solving the algebraic equation (3.20). It is clear that the approximation can be made as accurate as desired by just including more terms in the series. In addition to its computational simplicity, the method is amenable to error analysis.

Our objective is to determine an error bounded between the true and the approximate solutions to equation (3.16).

It is convenient to introduce the following notation: $G(\mathbf{r}, \mathbf{r}')$ denotes the approximate Green's function, and Ω denotes the finite set of subscripts $\{j\}$ corresponding to the terms that appear in the series representation of $G(\mathbf{r}, \mathbf{r}')$. That is,

$$G(\mathbf{r}, r) \triangleq \sum_{j \in \Omega} \frac{\psi_j(\mathbf{r})\psi_j(\mathbf{r}')}{b_0 + \lambda_j} \tag{3.22}$$

Also, if ϕ is a function in $L_2[V]$, then L, $\tilde{L}$, $\hat{L}$ and $\tilde{\hat{L}}$ denote the following transformations in $L_2[V]$:

$$L\phi \triangleq \int_V G(\mathbf{r}, \mathbf{r}')\phi(\mathbf{r}') \, dV' \tag{3.23}$$

$$\tilde{L}\phi \triangleq \int_V G(\mathbf{r}, \mathbf{r}')\phi(\mathbf{r}') \, dV' \tag{3.24}$$

$$\hat{L}\phi \triangleq L\hat{b}\phi = \int_V G(\mathbf{r}, r')\hat{b}(\mathbf{r}')\phi(\mathbf{r}') \, dV' \tag{3.25}$$

and

$$\tilde{\hat{L}}\phi \triangleq \tilde{L}\hat{b}\phi - \int_V G(\mathbf{r}, \mathbf{r}')\hat{b}(\mathbf{r}')\phi(\mathbf{r}') \, dV' \tag{3.26}$$

With these definitions, equation (3.16) and its approximate version can be written in operator form:

$$\psi_\phi(t) = Lf(t) - \hat{L}\psi_\phi(t) \tag{3.27}$$

and

$$\psi_\phi(t) = Lf(t) - \hat{L}\psi_\phi(t)\phi \tag{3.28}$$

where ψ_ϕ, ψ_ϕ, and f are shown as functions of t to emphasize their temporal dependency.

From equations (3.27) and (3.28) the error $\varepsilon(t) = \psi_\phi(t) - \psi_\phi(t)$ is given by

$$\varepsilon(t) = (L - \tilde{L})f(t) - [\hat{L}\psi_\phi(t) - \tilde{\hat{L}}\tilde{\psi}_\phi(t)] \tag{3.29}$$

which can be rewritten as

$$\varepsilon(t) = (L - \tilde{L})f(t) - \hat{L}\varepsilon(t) - (\hat{L} - \hat{L})\psi_\phi(t) \tag{3.30}$$

Hence, the $L_2[V]$ norm of $\varepsilon(t)$ is bounded above by

$$\|\varepsilon(t)\| \le \|(L - \tilde{L})f(t)\| + \|\hat{L}\| \, \|\varepsilon(t)\| + \|((\hat{L} - \hat{L})\psi_\phi(t)\| \tag{3.31}$$

and it follows that if $\hat{L}$ is a contraction operator then the error estimate is obtained in the form

$$\|\varepsilon(t)\| \le \frac{\|(L - L)f(t)\|}{1 - \|\hat{L}\|} + \frac{\|(\hat{L} - \tilde{L})\tilde{\psi}_\phi(t)\|}{1 - \|\hat{L}\|} \tag{3.32}$$

Given that the set $\{\psi_j\}$ is complete in the Hilbert space $L_2[V]$, every function ϕ in $L_2[V]$ has a Fourier series representation

$$\phi = \sum_j \psi_j \langle \psi_j, \phi \rangle \tag{3.33}$$

where $\langle \psi_j, \phi \rangle$ denotes the inner product in $L_2[V]$,

$$\langle \psi_j, \phi \rangle = \int_V \psi_j(\mathbf{r}) \bar{\phi}(\mathbf{r}) \, dV \tag{3.34}$$

and $\bar{\phi}$ is the complex conjugate of ϕ. Also, using the inner product notation, we can write

$$(L - \tilde{L})f(t) = \sum_{j \in \Omega} \frac{\psi_j \langle \psi_j, f(t) \rangle}{b_0 + \lambda_j} \tag{3.35}$$

and, given the orthonormality of the functions ψ_j,

$$(L - \tilde{L})f(t) = \sum_{j \in \Omega} \frac{\psi_j}{b_0 + \lambda_j} \left\langle \psi_j, \left(f(t) - \sum_{i \in \Omega} \psi_i \langle \psi_i, f(t) \rangle \right) \right\rangle \tag{3.36}$$

$$= (L - L)\left(f(t) - \sum_{i \in \Omega} \psi_i \langle \psi_i, f(t) \rangle \right) \tag{3.37}$$

Similarly, from the relations $\hat{L} = L\hat{b}$ and $\hat{L} = \tilde{L}\hat{b}$, it follows that

$$(\hat{L} - \tilde{L})\psi_\phi(t) = (L - \tilde{L})\left(\hat{b}\tilde{\psi}_\phi(t) - \sum_{i \in \Omega} \psi_i \langle \psi_i, \hat{b}\tilde{\psi}_\phi(t) \rangle \right) \tag{3.38}$$

Substituting equations (3.37) and (3.38) into equation (3.32), we obtain the error estimate

$$\|\varepsilon(t)\| \leq \frac{\|L - \tilde{L}\|}{1 - \|\hat{L}\|} \beta(t) \tag{3.39}$$

where

$$\beta(t) = \left\| f(t) - \sum_{i \in \Omega} \psi_i \langle \psi_i, f(t) \rangle \right\| + \left\| \hat{b}\tilde{\psi}_\phi(t) - \sum_{i \in \Omega} \psi_i \langle \psi_i, b\tilde{\psi}_\phi(t) \rangle \right\| \tag{3.40}$$

The norms of the operators L, $\tilde{L}$, $\hat{L}$, and $L - L$ are given by

$$\|\hat{L}\| = \max_j \frac{1}{|b_0 + \lambda_j|} \tag{3.41}$$

$$\|\tilde{L}\| = \max_{j \in \Omega} \frac{1}{|b_0 + \lambda_j|} \tag{3.42}$$

$$\|L - \tilde{L}\| = \max_{j \in \Omega} \frac{1}{|b_0 - \lambda_j|} \tag{3.43}$$

and

$$\|\hat{L}\| = \max_j \gamma_j^{1/2} \tag{3.44}$$

where $\{\gamma_j\}$ is the eigenvalue of $\hat{L}^*\hat{L}$ and $\hat{L}^*$ is the adjoint of $\hat{L}$. Often it is difficult to evaluate the norm of $\hat{L}$, and only an upper bound estimate can be obtained. A conservative estimate is

$$\|\hat{L}\| \leq \max_j \frac{1}{|b_0 + \lambda_j|} \max_{\mathbf{r}} |\hat{b}(\mathbf{r})| \tag{3.45}$$

Inequality (3.39) is a useful criterion to determine how many and which terms should be used to approximate the Green's function (3.15) by a series expansion. It is also clear from this inequality that b_0, in the decomposition of b as shown in equation (3.10), should be chosen so as to minimize the norm of $\hat{L}$.

3.2.3. Practical Example

The data for this example has been obtained by applying the modified-one-group approximation to the two-energy group data presented in Table 3.1. The modified-one-group data and the constants corresponding to the xenon and iodine dynamics are given in Table 3.2. The core configuration and the layout of zone controllers considered here is the same as in the example in Chapter 4. This is shown in Figure 4.2. The core is cylindrical and exhibits a degree of symmetry that corresponds to the point group D_{2h} of mathematical physics, which has eight unidimensional, irreducible representations.

Table 3.1. Two-Energy Neutronic Data for a Typical 1200 MW(th) Natural Uranium, Heavy-Water-Moderated, Pressurized-Tube, Nuclear Reactor

Infinite multiplication factor	$k_\infty = 1.404027$	
Resonance scape probability	$p = 0.89801$	
Neutron slowing down length	$L_s^2 = 134.3$	cm^2
Thermal diffusion length	$L^2 = 235.4$	cm^2
Fast neutron-group diffusion coeff.	$D_F = 1.3643$	cm
Slow neutron-group diffusion coeff.	$D_{th} = 1.2349$	cm
Fast neutron speed	$V_f = 10^7$	cm/sec
Slow neutron speed	$V = 3 \times 10^5$	cm/sec
Neutrons produced per fission	$v = 2.640$	
Core radius	$R = 400$	cm
Core length	$L = 600$	cm

Table 3.2. Data for the Modified One-Energy Model with Xenon and Iodine

Migration area	$M^2 = D/\sigma_a = 369.7 \text{ cm}^2$
Diffusion coefficient	$D = 1.2349 \text{ cm}$
Infinite multiplication factor	$k_\infty = M^2 v\sigma_f/D = 1.04027$
Xenon decay constant	$\lambda_x = 2.09 \times 10^{-5} \text{ sec}^{-1}$
Iodine decay constant	$\lambda_I = 2.87 \times 10^{-5} \text{ sec}^{-1}$
Iodine yield constant	$\gamma_I = 6.4 \times 10^{-2}$
Absorption macroscopic cross section of xenon	$\sigma = 1.22 \times 10^{-18} \text{ cm}^2$
Neutron flux distribution at steady state	$\phi_0 = a_0 \cos(\pi/L)zJ_0(\gamma_{01}r)$
	$a_0 = 0.1110 \times 10^{15}$
Core configuration	Cylindrical
Length of core	$L = 600 \text{ cm}$
Radius of core	$R = 400 \text{ cm}$
First root of $J_0(\gamma R)$	$\gamma_{01} = 6.01205 \times 20^{-5}$

The Laplacian modes for the cylindrical case $\{\psi_{nik}\}$ are ordered with the help of three subscripts. These modes are presented in Table 3.3, where these have been classified according to the irreducible representation of D_{2h}. The set of eigenvalues $\{\psi_{nik}\}$ corresponding to the Laplacian modes $\{\psi_{nik}\}$ are given in Table 3.3.

The parameter b_0 was chosen to be the constant value required to satisfy

$$\|b_0\| = \|b\| \tag{3.46}$$

where the norm is the usual $L_2[V]$ norm. The numerical values for $\|b\|$, b_0, and $\|\hat{b}\|$ are given in Table 3.4.

Two Laplacian modes per subsystem were considered to represent the Green's function (3.5), which in this case is of the form

$$G(r, \theta, z; r', \theta', z') = \sum_{q=1}^{8} G^{(q)}(r, \theta, z; r', \theta', z') \tag{3.47}$$

Table 3.3. The Norm of $L^{(q)}$, $\tilde{L}^{(q)}$, and $\hat{L}^{(q)}$

Subsystem (q)	$\Omega^{(q)} = \{(nik)\}$	$L^{(q)}$ and $\tilde{L}^{(q)}$	$L^{(q)} - \tilde{L}^{(q)}$	(Estimate) $\hat{L}^{(q)}$
1	$(1, 0, 1); (1, 2, 1)$	0.622012×10^5	0.723453×10^4	0.934
2	$(1, 2, 1); (1, 4, 1)$	0.513175×10^4	0.214191×10^4	0.7714×10^{-1}
3	$(1, 2, 1); (1, 4, 1)$	0.887950×10^4	0.256095×10^4	0.1335
4	$(1, 0, 1); (1, 2, 1)$	0.151127×10^5	0.453571×10^4	0.2272
5	$(1, 1, 1); (1, 3, 1)$	0.821102×10^4	0.296172×10^4	0.1234
6	$(1, 1, 1); (1, 3, 1)$	0.252902×10^5	0.391550×10^4	0.3801
7	$(1, 1, 1); (1, 3, 1)$	0.821102×10^4	0.296172×10^4	0.1234
8	$(1, 1, 1); (1, 3, 1)$	0.252902×10^5	0.391550×10^4	0.3801

Table 3.4. The Laplacian Eigenvalues

Subsystem q	Eigenvalue $\lambda_{nik}^{(q)}$	
1	$(2n-1)^2(\pi/L)^2 + \gamma_{ik}^2$	$i = 0, 2, 4, \ldots$
2	$(2n)^2(\pi/L)^2 + \gamma_{ik}^2$	$i = 2, 4, 6, \ldots$
3	$(2n-1)^2(\pi/L) + \gamma_{ik}^2$	$i = 2, 4, 6, \ldots$
4	$(2n)^2(\pi/L)^2 + \gamma_{ik}^2$	$i = 0, 2, 4, \ldots$
5	$(2n)^2(\pi/L)^2 + \gamma_{ik}^2$	$i = 1, 3, 5, \ldots$
6	$(2n-1)^2(\pi/L)^2 + \gamma_{ik}^2$	$i = 1, 3, 5, \ldots$
7	$(2n)^2(\pi/L)^2 + \gamma_{ik}^2$	$i = 1, 3, 5, \ldots$
8	$(2n-1)^2(\pi/L)^2 + \gamma_{ik}^2$	$i = 1, 3, 5, \ldots$

where

$$G^{(q)}(r, \theta, z; r', \theta', z') = \sum_{(nik)\in\Omega^{(q)}} \frac{\psi_{nik}^{(q)}(r, \theta, z)\psi_{nik}^{(q)}(r', \theta', z')}{b_0 + \lambda_{nik}^{(q)}} \quad (3.48)$$

where the superscript (q) corresponds to the qth subsystem. The modes in each subsystem q were selected to minimize the norm $\|L^{(q)} - \tilde{L}^{(q)}\|$.

Table 3.5 shows the subscripts corresponding to the selected functions, and the associated norms of $L^{(q)}$, $L^{(q)} - \tilde{L}^{(q)}$, and the estimates for $\|\hat{L}^{(q)}\|$. The estimates $\|\hat{L}^{(q)}\|$ were evaluated from

$$\|\hat{L}^{(q)}\| \leq \|L^{(q)}\| b' \quad (3.49)$$

where b' is an effective constant such that

$$\|b'\| = \|\hat{b}\| \quad (3.50)$$

The maximum values of $\hat{b}$ occur near the center of the core and at the extrapolated boundary. Since the flux is zero at the extrapolated boundary, and since the contribution to the norm of $\|\hat{b}_\phi\|$ by a neighborhood of small radius and located in the center of the core is negligible, it can be seen that for more flux shapes of practical interest the estimate in equation (3.49) is indeed valid.

The error estimates in terms of the function f and the approximate solution ψ_ϕ are given by

$$\|\varepsilon(t)\| = \left\| \sum_{q=1}^{8} \varepsilon^{(q)}(t) \right\| \leq \sum_{q=1}^{8} \frac{\|L^{(q)} - \tilde{L}^{(q)}\|}{1 - \|\hat{L}^{(q)}\|} \beta^{(q)}(t) \quad (3.51)$$

Table 3.5. The Parameter b_0 and the $L_2[V]$ Norms
of b and $\hat{b}$

$\|b\|$	$b_0 = \|b\|/V^{1/2}$	$\|b - b_0\|$
0.138299×10^1	0.79636×10^{-4}	0.261035

where

$$\beta^q(t) = \left\| f^{(q)} - \sum_{nik \in \Omega^{(q)}} \psi_{nik}^{(q)} f_{nik}^{(q)} \right\| + \left\| \hat{b}\tilde{\phi}_\phi^{(q)} - \sum_{nik \in \Omega^{(q)}} \phi_{nik}^{(q)} g_{nik}^{(q)} \right\| \quad (3.52)$$

$$f_{nik}^{(q)} = \int_{-1/2}^{1/2} \int_0^{2\pi} \int_0^R \psi_{nik}^{(q)}(r, \theta, z) f(r, \theta, z, t) r \, dr \, d\theta \, dz \quad (3.53)$$

$$g_{nik}^{(q)} = \int_{-1/2}^{1/2} \int_0^{2\pi} \int_0^R \psi_{nik}^{(q)}(r, \theta, z) \hat{b}(r, \theta, z) \tilde{\psi}_\phi^{(q)}(r, \theta, z, t) r \, dr \, d\theta \, dz \quad (3.54)$$

The method of degenerate kernels was then applied to obtain the approximate solution $\tilde{\psi}_\phi$ as follows. The numerical values of the parameters $a_{nik}^{(q)}$, $\xi_{nik}^{(q)}$, and $d_{nik}^{(q)}$ are presented in Table 3.6.

$$\psi_\phi(r, \theta, z, t) = \sum_{q=1}^{8} \tilde{\psi}_\phi^{(q)}(r, \theta, z, t) \quad (3.55)$$

$$\tilde{\psi}_\phi^{(q)}(r, \theta, z, t) = \sum_{(nik) \in \Omega^{(q)}} a_{nik}^{(q)} \psi_{nik}^{(q)}(r, \theta, z)$$

$$\times \int_{-L/2}^{L/2} \int_0^{2\pi} \int_0^R \psi_{nik}^{(q)}(r', \theta', z') f(r', \theta', z', t) r' \, dr' \, d\theta' \, dz' \quad (3.56)$$

Table 3.6. The Expansion Coefficients $a_{nik}^{(q)}$

Subsystem (q)	(nik)	$\xi_{nik}^{(q)}$	$d_{nik}^{(q)}$	$a_{nik}^{(q)}$
1	1 0 1	0.622012×10^5	3.8955	1.2706×10^4
	1 2 1	-0.887950×10^4	-5.9530×10^{-1}	-2.1941×10^4
2	1 2 1	-0.513175×10^4	-3.6018×10^{-1}	-8.0206×10^3
	1 4 1	-0.256466×10^4	-1.8795×10^{-1}	-3.1583×10^3
3	1 2 1	-0.8887950×10^4	-5.9530×10^{-1}	-2.1941×10^4
	1 4 1	-0.325026×10^4	-2.2745×10^{-1}	-4.2072×10^3
4	1 0 1	-0.151127×10^5	-9.8827×10^{-1}	-1.2888×10^6
	1 2 1	-0.513175×10^4	-3.6018×10^{-1}	-8.0206×10^3
5	1 1 1	-0.821102×10^4	-5.5932×10^{-1}	-1.8633×10^4
	1 3 1	-0.351564×10^4	-2.5267×10^{-1}	-4.7043×10^3
6	1 1 1	-0.252902×10^5	-1.6469	3.9097×10^4
	1 3 1	-0.494566×10^4	-3.3942×10^{-1}	-7.4869×10^3
7	1 1 1	-0.821102×10^4	-5.5932×10^{-1}	-1.8633×10^4
	1 3 1	-0.351564×10^4	-2.5267×10^{-1}	-4.7043×10^3
8	1 1 1	-0.252902×10^5	-1.6469	3.9097×10^4
	1 3 1	-0.494566×10^4	-3.3942×10^{-1}	-7.4869×10^3

$$a_{nik}^{(q)} = \frac{\xi_{nik}^{(q)}}{1 + d_{nik}^{(q)}} \tag{3.57}$$

$$\xi_{nik}^{(q)} = \frac{1}{b_0 + \lambda_{nik}^{(q)}} \tag{3.58}$$

$$d_{nik}^{(q)} = \xi_{nik}^{(q)} \int_{-L/2}^{L/2} \int_0^{2\pi} \int_0^R \psi_{nik}^{(q)}(r', \theta', z')\hat{b}(r', \theta', z')$$

$$\times \psi_{nik}^{(q)}(r', \theta', z')r' \, dr' \, d\theta' \, dz' \tag{3.59}$$

3.3. The Model Expansion

From the discussion of the previous section it is clear that the neutron flux ψ_ϕ can be expressed in terms of the function f in operator form

$$\psi_\phi = Ff \tag{3.60}$$

where F is the operator (or an approximation to) $[I + \hat{L}]^{-1}L$. In the particular case of the example, equation (3.60) corresponds to equation (3.56).

Substituting equation (3.3) in (3.60), we find

$$\psi_\phi = \frac{\sigma}{D} F\phi_0\psi_x + F\hat{u} \tag{3.61}$$

Eliminating the flux from equations (3.5) and (3.6), we obtain the dynamic equations for xenon and iodine:

$$\frac{\partial \psi_I}{\partial t} = -\lambda_I\psi_I + \gamma_I\Sigma_f\frac{\sigma}{D} F\phi_0\psi_x + \gamma_I\Sigma_f F\hat{u} \tag{3.62}$$

$$\frac{\partial \psi_x}{\partial t} = \lambda_I\psi_I - \left(\lambda_x + \sigma\phi_0 + \frac{\sigma^2}{D} x_0 F\phi_0\right)\psi_x - \sigma x_0 F\hat{u} \tag{3.63}$$

with initial conditions

$$\psi_I(\mathbf{r}, t_0) = \psi_{I0}(\mathbf{r}) \tag{3.64}$$

$$\phi_r(\mathbf{r}, t_0) = \psi_{x0}(\mathbf{r}) \tag{3.65}$$

Equations (3.62) and (3.63) can be written as an operator equation in the product space $L_2[V] \times L_2[V]$:

$$\frac{\partial}{\partial t} \psi = \mathbf{A}\psi + \mathbf{z} \tag{3.66}$$

where

$$\psi = \text{col}[\psi_I, \psi_x] \tag{3.67}$$

$$z = \text{col}[z_I, z_x] \tag{3.68}$$

$$z_I = \gamma_I \Sigma_f F \hat{u} \tag{3.69}$$

$$z_x = -x_0 \sigma F \hat{u} \tag{3.70}$$

and

$$A = \begin{bmatrix} -\lambda_I & \gamma_I \Sigma_f \dfrac{\sigma}{D} f\phi_0 \\[2mm] \lambda_I & -\left(\lambda_x + \sigma\phi_0 + x_0 \dfrac{\sigma^2}{D} F\phi_0\right) \end{bmatrix} \tag{3.71}$$

Note that the linear analysis of xenon spatial oscillation could be carried out by computing the eigenvalues of the operator A. This procedure would be equivalent to studying xenon stability by means of Kaplan's natural modes. Several difficult theoretical and computational problems, however, make this approach impractical. A simple computation will show that the operator A is not compact, non-self-adjoint, and furthermore, not even normal. The theoretical implication of this is that nothing can be said about the existence and completeness of the eigenvalues of A. Even if it is assumed that the eigenvalue spectrum of A is discrete, the numerical computation of the eigenfunctions and their corresponding eigenvalues is difficult. Successful computations have been reported only in the case of a slab reactor. Therefore, several methods have been proposed to obtain approximations to the natural modes. Of these, probably the best known are the lambda and mu methods developed by Stacey (Refs. 3.12–3.14). Roughly speaking, these methods consist of Laplace-transforming equation (3.64) and manipulating the resulting equations in a way that allows the use of conventional, static diffusion codes.

Our purpose here is not to perform a stability analysis but to find an approximate solution to equation (3.64) in terms of the control variable $\hat{u}$. It is obvious, however, that the two problems are interrelated.

The proposed method consists of expanding the iodine and xenon concentrations in terms of a finite number of Laplacian modes $\{\psi_j\}$ and obtaining a weak solution to equations (3.63) and (3.64) in the form of an integral equation. Let ΩI denote the finite set of subscripts corresponding to the modes in the finite expansion, and let ψ_I and ψ_x denote the approximate solutions

$$\tilde{\psi}_I = \sum_{i \in \Omega I} \psi_i W_{I_i} \tag{3.72}$$

$$\tilde{\phi}_x = \sum_{i \in \Omega I} \psi_i W_{x_i} \tag{3.73}$$

where the coefficients W_{I_i} and W_{x_i} satisfy the system of ordinary differential equations

$$\frac{d}{dt} W_{I_i}(t) = -\lambda_I W_{I_i}(t) - \sum_{j \in \Omega I} \left\langle \psi_i, \gamma_I \Sigma_f \frac{\sigma}{D} F\phi_0 \psi_j \right\rangle W_{x_j}(t) + \langle \phi_i, z_I(t) \rangle \tag{3.74}$$

and

$$\frac{d}{dt} W_{x_i}(t) = \lambda_I W_{I_i}(t) - \sum_{j \in \Omega I} \left\langle \psi_j, \left(\lambda_x + \sigma\phi_0 + \frac{\sigma^2}{D} x_0 F\phi_0 \right) \psi_j \right\rangle W_{x_j}(t)$$
$$+ \langle \psi_j, z_x(t) \rangle \tag{3.75}$$

with initial conditions

$$W_{I_i}(t_0) = \langle \psi_i, \psi_I(t_0) \rangle \tag{3.76}$$

$$W_{x_j}(t_0) = \langle \psi_i, \psi_x(t_0) \rangle \tag{3.77}$$

Multiplying equations (3.74) and (3.75) by ψ_i and summing over the set ΩI, we see that ψ_I and ψ_x satisfy the equations

$$\frac{\partial}{\partial t} \tilde{\psi}_I = -\lambda_I \tilde{\psi}_L + \sum_{i,j \in \Omega I} \psi_i \left\langle \psi_i \gamma_I \Sigma_f \frac{\sigma}{D} F\phi_0 \psi_j \right\rangle \langle \psi_j, \tilde{\psi}_x \rangle + \sum_{i \in \Omega I} \psi_i \langle \tilde{\psi}_i, z_I \rangle \tag{3.78}$$

$$\frac{\partial}{\partial t} \tilde{\psi}_x = \lambda_I \tilde{\psi}_x - \sum_{i,j \in \Omega I} \psi_i \left\langle \psi_i, \left(\lambda + \sigma\phi_0 + x_0 \frac{\sigma^2}{D} F\phi_0 \right) \psi_j \right\rangle \langle \psi_j, \tilde{\psi}_x \rangle$$
$$+ \sum_{i \in \Omega I} \psi_i \langle \psi_i, z_x \rangle \tag{3.79}$$

To estimate the approximating error, it is convenient to combine equations (3.78) and (3.79) in an operator equation similar to equation (3.64). To this end, we introduce the following definitions:

$$\tilde{z}_1 = \sum_{i \in \Omega I} \psi_i \langle \psi_i, z_I \rangle \tag{3.80}$$

$$\tilde{z}_2 = \sum_{i \in \Omega I} \psi_i \langle \psi_i, z_x \rangle \tag{3.81}$$

Also, the operators $\tilde{A}_1$ and $\tilde{A}_2$ with range and domain in $L_2[V]$ are defined by

$$\tilde{A}_1 \tilde{\psi}_x = \sum_{i,j \in \Omega I} \psi_i \left\langle \psi_i, \gamma_I \Sigma_f + \frac{\sigma}{D} F\phi_0 \psi_j \right\rangle \langle \psi_j, \tilde{\psi}_x \rangle \tag{3.82}$$

and

$$\tilde{A}_2 \tilde{\psi}_x = \sum_{i,j \in \Omega I} \psi_i \left\langle \psi_i, \left(\lambda_x + \sigma\phi_0 + \frac{\sigma^2}{D} X_0 F\phi_0 \right) \psi_j \right\rangle \langle \psi_j, \psi_x \rangle \tag{3.83}$$

Finally, let

$$\tilde{\psi} = \text{col}[\tilde{\psi}_I, \tilde{\psi}_x] \tag{3.84}$$

$$\tilde{z} = \text{col}[\tilde{z}_I, \tilde{z}_x] \tag{3.85}$$

We can now write equations (3.78) and (3.79) in the form

$$\frac{\partial}{\partial t} \tilde{\psi} = \tilde{A}\tilde{\psi} + \tilde{z} \tag{3.86}$$

where

$$A = \begin{bmatrix} \lambda_I & \tilde{A}_1 \\ -\lambda_I & \tilde{A}_2 \end{bmatrix} \tag{3.87}$$

In view of equations (3.63) and (3.86), it follows that the error $\varepsilon = \psi - \tilde{\psi}$ between the true solution to equation (3.64) and $\tilde{\psi}$ satisfies

$$\frac{\partial}{\partial t} \varepsilon = A\psi - \tilde{A}\tilde{\psi} + z - \tilde{z} = A\varepsilon + (A - \tilde{A})\tilde{\psi} + z - \tilde{z} \tag{3.88}$$

with initial condition

$$\varepsilon(t_0) = \psi(t_0) - \tilde{\psi}(t_0) \tag{3.89}$$

If A is expressed as the addition of two matrix operators in $L_2[V] \times L_2[V]$,

$$A = A_0 + \hat{A} \tag{3.90}$$

where A_0 is a two-dimensional matrix with constant coefficients, then equation (3.88) could be written in the form

$$\frac{\partial}{\partial t} \varepsilon = A_0\varepsilon + \hat{A}\varepsilon + h \tag{3.91}$$

where

$$h = (A - \tilde{A})\tilde{\psi} + z - \tilde{z} \tag{3.92}$$

Then using the familiar convolution integral, we have

$$\varepsilon(t) = g(t) + \int_0^t \phi_0(t, \tau)\hat{A}\varepsilon(\tau) \, d\tau \tag{3.93}$$

where $\phi_0(t, \tau)$ is the fundamental matrix associated with A_0 and

$$g(t) = \phi_0(t, t_0)\varepsilon(t_0) + \int_{t_0}^t \phi_0(t, \tau)h(\tau) \, d\tau \tag{3.94}$$

Equation (3.93) has the familiar form of the nonhomogeneous Volterra equation of the second kind in the product space $L_2[V] \times L_2[V]$. It is well known that a unique solution to (3.93) exists, provided that the operator $\phi_0(t, \tau)\hat{A}$ is bounded in $L_2[V] \times L_2[V]$ for all finite values of $t - \tau$. Furthermore, the solution can be expressed in the Neumann series (contraction mapping series)

$$\varepsilon(T) = g(t) + \sum_{n=1}^{\infty} \int_{t_0}^{t} \mathbf{K}_n(t, \tau)g(\tau)\, d\tau \qquad (3.95)$$

where

$$\mathbf{K}_1(t, \tau) = \phi_0(t, \tau)\hat{A} \qquad (3.96)$$

and

$$\mathbf{K}_n(t, \tau) = \int_{\tau}^{t} \phi_0(t, s)\hat{A}\mathbf{K}_{n-1}(s, \tau)\, ds \qquad (3.97)$$

Consider a real, bounded, and positive number M such that, for all $t - \tau$,

$$\|\mathbf{K}_1(t, \tau)g(\tau)\|_2 \le M\|g(\tau)\|_2 \qquad (3.98)$$

where $\|\cdot\|_2$ denotes the norm in $L_2[V] \times L_2[V]$ inherited from the inner product $\langle\ ,\ \rangle_2$,

$$\langle g, g\rangle_2 = \langle g_1, g_1\rangle + \langle g_2, g_2\rangle \qquad (3.99)$$

and let

$$M_g = \max_t \|g(t)\|_2 \qquad (3.100)$$

Then it follows that

$$\|\mathbf{K}_2(t, \tau)g(\tau)\|_2 \le \int_{\tau}^{t} \|K_1(t, s)\|_2 MM_g\, ds = (t - \tau)M^2 M_g \qquad (3.101)$$

and, by induction,

$$\|\mathbf{K}_n(t, \tau)g(\tau)\|_2 \le \frac{(t - \tau)^{n-1}}{(n - 1)!} M^n M_g \qquad (3.102)$$

which tends to zero as $n \to \infty$. Using equation (3.102) in equation (3.95), we have that

$$\|\varepsilon(t)\|_2 \le \|g(t)\|_2 + \sum_{n=1}^{\infty} \int_{t_0}^{t} \|\mathbf{K}_n(t, \tau)g(\tau)\|\, d\tau$$

$$\le \left[1 + \sum_{n=1}^{\infty} \frac{(t - t_0)^n}{n!} M^n\right] M_g$$

$$= \exp(t - t_0)M_{M_g} \qquad (3.103)$$

Even though this error estimate is very conservative, it can, together with equations (3.92) and (3.94), help to determine how good the approximation ψ is. In particular, equation (3.92) is a good criterion for checking the adequacy of the approximation. Note also that the sharpest estimate obtainable by this approach would be determined by choosing A_0 so as to minimize the bound M in eq. (3.98).

3.3.1. The Functional Relation

Using the convolution integral, we obtain the solution to the system of equations (3.74) and (3.75) in the form

$$W_{I_i}(t) = \sum_{j \in \Omega I} (\phi_{II_{ij}}(t, t_0) W_{I_j}(t_0) + \phi_{IX_{ij}}(t, t_0) W_{x_j}(t_0))$$

$$+ \sum_{j \in \Omega I} \int_{t_0}^{t} (\phi_{II_{ij}}(t, \tau) \langle \psi_j, z_I(\tau) \rangle + \phi_{IX_{ij}}(t, \tau) \langle \psi_j, z_x(\tau) \rangle) \, d\tau \quad (3.104)$$

and

$$W_{x_i}(t) = \sum_{j \in \Omega I} (\phi_{XI_{ij}}(t, t_0) W_{I_j}(t_0) + \phi_{XX_{ij}}(t, t_0) W_{x_j}(t_0))$$

$$+ \sum_{j \in \Omega I} \int_{t_0}^{t} (\phi_{XI_{ij}}(t, \tau) \langle \psi_j, z_I(\tau) \rangle + \phi_{XX_{ij}}(t, \tau) \langle \psi_j, z_x(\tau) \rangle) \, dt \quad (3.105)$$

where the time-dependent functions $\phi_{II_{ij}}$, $\phi_{IX_{ij}}$, $\phi_{XI_{ij}}$, and $\phi_{XX_{ij}}$ are the components of the matrices Φ_{II}, Φ_{IX}, Φ_{XI}, Φ_{XX} that form the fundamental matrix

$$\Phi = \begin{bmatrix} \Phi_{II} & \Phi_{IX} \\ \Phi_{II} & \Phi_{XX} \end{bmatrix} \quad (3.106)$$

associated with

$$\begin{bmatrix} A_{II} & A_{IX} \\ A_{XI} & A_{XX} \end{bmatrix} \quad (3.107)$$

where A_{II}, A_{IX}, A_{XI}, and A_{XX} are square matrices of dimension equal to the number of subscripts in ΩI and

$$A_{II} = \text{diag}[-\lambda_I] \quad (3.108)$$

$$A_{XI} = -A_{II} \quad (3.109)$$

Finally, A_{IX} and A_{XX} have components $a_{IX_{ij}}$ and $a_{XX_{ij}}$ defined by

$$a_{IX_{ij}} = \left\langle \psi_i, \gamma_i \Sigma_f \frac{\sigma}{D} F \phi_0 \psi_j \right\rangle \quad (3.110)$$

and

$$a_{XX_{ij}} = -\left\langle \psi_i, \left(\lambda_x + \sigma\phi_0 + \frac{\sigma^2}{D} X_0 F\phi_0 \right) \psi_j \right\rangle \tag{3.111}$$

Substituting equations (3.69) and (3.70) into equation (3.104), we obtain the coefficients of the xenon expansion in terms of the control variable $\hat{u}$:

$$W_{x_i}(t) = \sum_{j \in \Omega I} (\phi_{XI_{ij}}(t, t_0)\langle\psi_j, \psi_I(t_0)\rangle + \phi_{XX_{ij}}(t, t_0)\langle\psi_j, \psi_x(t_0)\rangle)$$

$$+ \sum_{j \in \Omega I} \int_{t_0}^{t} (\phi_{XI_{ij}}(t, \tau)\langle\psi_j, \gamma_I \Sigma_f F\hat{u}(\tau)\rangle$$

$$- \phi_{XX_{ij}}(t, \tau)\langle\psi_j, X_0 \sigma F\hat{u}(\tau)\rangle)\, d\tau \tag{3.112}$$

From equation (3.61), repeated here,

$$\psi_\phi = \frac{\sigma}{D} F\phi_0\psi_x + F\hat{u} \tag{3.113}$$

and since $\psi_x = \sum_{i \in \Omega I} \psi_i W_{x_i}$, we obtain the neutron flux in terms of the control variable U:

$$\psi_\phi(t) = \sum_{i \in \Omega I} \frac{\sigma}{D} F\phi_0\psi_i W_{x_i}(t) + F\hat{u}(t) \tag{3.114}$$

where for simplicity only the temporal dependence is explicitly shown.

Substituting equation (3.112) into equation (3.114), we get the desired functional relation:

$$\psi_\phi = \Delta + F\hat{u} + F_1\hat{u} \tag{3.115}$$

where Δ is a functional in $L_2(V)$ defined by

$$\Delta(t) = \sum_{i \in \Omega I} \frac{\sigma}{D} F\phi_0\psi_i \sum_{j \in \Omega I} (\phi_{XI_{ij}}(t, t_0)\langle\psi_j, \psi_I(t_0)\rangle$$

$$+ \phi_{XX_{ij}}(t, t_0)\langle\psi_j, \psi_x(t_0)\rangle) \tag{3.116}$$

and F_1 is the transformation from the space of control functions into $L_2[V]$:

$$F_1\hat{u}(t) = \sum_{i,j \in \Omega I} \frac{\sigma}{D} F\phi_0\psi_i \int_{t_0}^{t} (\phi_{XI_{ij}}(t, \tau)\langle\psi_j, \gamma_I \Sigma_f F\hat{u}(\tau)\rangle$$

$$- \phi_{XX_{ij}}(t, \tau)\langle\psi_j, X_0 \sigma F\hat{u}(\tau)\rangle)\, d\tau \tag{3.117}$$

3.3.2. Practical Example

By the example of Section 3.2.3, the matrices $A_{II}^{(q)}$, $A_{IX}^{(q)}$, $A_{XI}^{(q)}$, and $A_{XX}^{(q)}$ corresponding to the qth subsystem are diagonal and two-dimensional. The diagonal entries in $A_{IX}^{(q)}$ and $A_{XX}^{(q)}$,

$$a_{IX_{(nik)}}^{(q)} = \left\langle \psi_{nik}^{(q)}, \gamma_I \Sigma_f \frac{\sigma}{D} F \phi_0 \psi_{nik}^{(q)} \right\rangle \tag{3.118}$$

and

$$a_{XX_{(nik)}}^{(q)} = -\left\langle \psi_{nik}^{(q)}, \left(\lambda_x + \sigma \phi_0 + \frac{\sigma^2}{D} X_0 F \phi_0 \right) \psi_{nik}^{(q)} \right\rangle \tag{3.119}$$

are given in Table 3.7, in which the eigenvalues $p_{1_{(nik)}}^{(q)}$ and $p_{2_{(nik)}}^{(q)}$ of the augmented matrix

$$\begin{bmatrix} A_{II}^{(q)} & A_{IX}^{(q)} \\ A_{XI}^{(q)} & A_{XX}^{(q)} \end{bmatrix} \tag{3.120}$$

are also shown.

It can be seen from the eigenvalues of Table 3.7 that, although all the subsystems exhibit oscillatory behavior, all but one of the subsystems are stable. The mode $\psi_{101}^{(4)}$ is unstable. Of all the stable modes, $\psi_{111}^{(6)}$ exhibits the slowest transient response with a time constant of 47.13 h.

The neutron flux ψ_ϕ is obtained in the form

$$\psi_\phi = \sum_{q=1}^{8} \psi_\phi^{(q)} \tag{3.121}$$

where

$$\psi_\phi^{(q)} = \Delta^{(q)} + F^{(q)} \hat{u}^{(q)} + F_1^{(q)} \hat{u}^{(q)} \tag{3.122}$$

Expanding the control variables $\hat{u}^{(q)}$ in terms of the symmetry-adapted linear combinations $\{\rho_k^{(q)}\} = k = 1, C_q$ of Section 3.4,

$$\hat{u}^{(q)}(t) = \sum_{k=1}^{C_q} \phi_0 \rho_k^{(q)} y_k^{(q)}(t) \tag{3.123}$$

and using the spectral representation of $F^{(q)}$ of equation (3.53), we have terms $\Delta^{(q)}$, $F^{(q)} \hat{u}^{(q)}$, and $F_1^{(q)} \hat{u}^{(q)}$:

$$\Delta^{(q)}(t) = \sum_{(nik) \in \Omega^{(q)}} \frac{\sigma}{D} \psi_{nik}^{(a)} A_{nik}^{(q)} \langle \psi_{nik}^{(q)}, \phi_0 \psi_{nik}^{(q)} \rangle \Delta_{nik}^{(q)}(t) \tag{3.124}$$

where

$$\Delta_{nik}^{(q)}(t) = \phi_{XI_{(nik)}}^{(q)}(t, t_0) \langle \psi_{nik}^{(q)}, \psi_I(t_0) \rangle + \phi_{XX_{(nik)}}^{(q)}(t, t_0) \langle \psi_{nik}^{(q)}, \psi_X(t_0) \rangle \tag{3.125}$$

The time-dependent functions ϕ_{XI} and ϕ_{XX} are defined by

$$\phi_{XI_{nik}}^{(q)}(t - \tau) = \frac{-\lambda_I}{p_{1_{nik}}^{(q)} - p_{2_{nik}}^{(q)}} [\exp p_{1_{nik}}^{(q)}(t - \tau) - \exp p_{2_{nik}}^{(q)}(t - \tau)] \tag{3.126}$$

Table 3.7. System Eigenvalues

Subsystem	nik	$a_{1x_{mk}}^{(q)}$	$a_{xx_{mk}}^{(q)}$	Eigenvalues	
				$p_{1_{mk}}^{(q)}$	$p_{2_{mk}}^{(q)}$
1	101	6.2988×10^{-5}	-1.5289×10^{-4}	-1.5538×10^{-1}	-1.6605×10^{-5}
	121	-7.5823×10^{-5}	-2.5597×10^{-5}	$(-2.7149+j4.6623)\times10^{-5}$	$(2.7129-j4.6623)\times10^{-5}$
2	121	-2.2174×10^{-5}	-5.2864×10^{-5}	$(4.0773+j2.2150)\times10^{-5}$	$(-4.0773-j2.2150)\times10^{-5}$
	141	-6.9615×10^{-5}	-5.33696×10^{-5}	$(4.1197+j4.2916)\times10^{-5}$	$(-4.1197-j4.2916)\times10^{-5}$
3	121	-7.5823×10^{-5}	-2.5597×10^{-5}	$(-2.7149+j4.6623)\times10^{-5}$	$(-2.7149-j4.6623)\times10^{-5}$
	141	-1.1593×10^{-5}	-5.9523×10^{-5}	$(-4.4112+j0.9757)\times10^{-5}$	$(-4.4112+j0.9757)\times10^{-5}$
4	101	-5.1112×10^{-3}	3.6426×10^{-3}	3.6022×10^{-3}	1.1701×10^{-5}
	121	-2.2174×10^{-5}	-5.2846×10^{-5}	$(-4.0773+j2.2150)\times10^{-5}$	$(-4.0773-j2.2150)\times10^{-5}$
5	111	-6.0093×10^{-5}	-3.3855×10^{-5}	$(-3.1278+j4.1449)\times10^{-5}$	$(-3.1278-j4.14449)\times10^{-5}$
	131	-1.1497×10^{-5}	-5.4727×10^{-5}	$(-4.1714-j1.2673)\times10^{-5}$	$(-4.1714-j1.2673)\times10^{-5}$
6	111	1.5761×10^{-4}	-2.0423×10^{-4}	-5.8933×10^{-6}	-2.2704×10^{-4}
	131	-2.2872×10^{-5}	-5.6661×10^{-5}	$(-4.2681+j2.1470)\times10^{-5}$	$(-4.2681-j2.1470)\times10^{-5}$
7	111	-6.0093×10^{-5}	-3.3855×10^{-5}	$(-3.1278+j4.1449)\times10^{-5}$	$(-3.1278+j4.1449)\times10^{-5}$
	131	-1.1497×10^{-5}	-5.4727×10^{-5}	$(-4.1714-j1.2673)\times10^{-5}$	$(-4.1714-j1.2673)\times10^{-5}$
8	111	1.57661×10^{-4}	-2.0423×10^{-4}	-5.8933×10^{-6}	-2.2704×10^{-4}
	131	-2.2872×10^{-5}	-5.6661×10^{-5}	$(-4.2681+j2.1470)\times10^{-5}$	$(-4.2681-j2.1470)\times10^{-5}$

and

$$\phi_{XX_{nik}}^{(q)}(t - \tau) = \frac{1}{p_{1_{nik}}^{(q)} - p_{2_{nik}}^{(q)}} [(p_{1_{nik}}^{(q)} + \lambda_I) \exp p_{1_{nik}}^{(q)}(t - \tau)$$
$$- (p_{2_{nik}}^{(q)} + \lambda_I) \exp p_{2_{nik}}^{(q)}(t - \tau)] \qquad (3.127)$$

if the eigenvalues are real, or

$$\phi_{XI_{nik}}^{(q)}(t - \tau) = \frac{-\lambda_I}{IM[p_{1_{nik}}^{(q)}]} \exp RE[p_{1_{nik}}^{(q)}](t - \tau) \sin(IM[p_{1_{nik}}^{(q)}](t - \tau))$$
$$(3.128)$$

and

$$\phi_{XX_{nik}}^{(q)}(t - \tau) = h_{nik}^{(q)} \exp RE[p_{1_{nik}}^{(q)}](t - \tau) \sin(IM[p_{1_{nik}}^{(q)}(t - \tau) + \phi_{nik}^{(q)})$$
$$(3.129)$$

where

$$h_{nik}^{(q)} = \frac{((RE[p_{1_{nik}}^{(q)}] + \lambda_I)^2 + IM[p_{1_{nik}}^{(q)}]^2)^{1/2}}{IM[p_{1_{nik}}^{(q)}]} \qquad (3.130)$$

and

$$\phi_{nik}^{(q)} = \tan^{-1}\left(\frac{IM[p_{1_{nik}}^{(q)}]}{RE[p_{1_{nik}}^{(q)} + \lambda_I]}\right) \qquad (3.131)$$

Also,

$$F^{(q)}\hat{u}^{(q)} = \sum_{m=1}^{C_q} \sum_{(nik)\in\Omega^{(q)}} \psi_{nik}^{(q)} a_{nik}^{(q)} \langle \psi_{nik}^{(q)}, \phi_0 \rho_m^{(q)} \rangle y_m^{(q)}(r) \qquad (3.132)$$

Finally, we obtain

$$F_1^{(q)}\hat{u}^{(q)} = \sum_{(nik)\in\Omega^{(q)}} \psi_{nik}^{(q)} a_{nik}^{(q)} \frac{\sigma}{D} \langle \psi_{nik}^{(q)}, \phi_0 \psi_{nik}^{(q)} \rangle f_{1_{nik}}^{(q)}(t) \qquad (3.133)$$

and

$$f_{1_{nik}}^{(q)}(t) = \sum_{m=1}^{C_q} \int_{t_0}^{t} (\phi_{XI_{nik}}^{(q)}(t - \tau) \langle \psi_{nik}^{(q)}, \phi_0 \rho_m^{(q)} \rangle a_{nik}^{(q)} \gamma_I \Sigma_f$$
$$- \phi_{XX_{nik}}^{(q)}(t - \tau) \langle \psi_{nik}^{(q)}, X_0 \psi_{nik}^{(q)} \rangle a_{nik}^{(q)} \langle \psi_{nik}^{(q)}, \phi_0 \rho_m^{(q)} \rangle \sigma) y_m^{(q)}(\tau) \, d\tau$$
$$(3.134)$$

where the coefficients are given in Tables 3.7, 3.8, and 3.9.

Table 3.8. Inner Product Values

Subsystem q	nik	$\langle \psi_{nik}^{(q)}, \phi_0 \psi_{nik}^{(q)} \rangle \times 10^{-14}$	$\langle \psi_{nik}^{(q)}, X_0 \psi_{nik}^{(q)} \rangle \times 10^{-14}$
1	101	0.6812	0.4686
	121	0.4784	0.4239
2	121	0.3799	0.3921
	141	0.3029	0.3607
3	121	0.4784	0.4239
	141	0.3786	0.3942
4	101	0.5449	0.4406
	121	0.3799	0.3921
5	111	0.4431	0.4131
	131	0.3358	0.3751
6	111	0.5539	0.4434
	131	0.4198	0.4079
7	111	0.4431	0.4131
	131	0.3358	0.3751
8	111	0.5539	0.4434
	131	0.4198	0.4079

3.4. Model Decomposition Techniques

The discussion in the previous section on the application of optimal control techniques applies to distributed parameter models in general. However, practical implementation of the derived theoretical results to more realistic reactor models, involving, for example, geometrical configurations in (2) or (3) dimensions and including the dynamics of fission products such as xenon and iodine, is possible only if we use efficient methods for treating the mathematical models and the resulting conditions for optimality.

This section is devoted to the problem of decoupling or reducing either the mathematical reactor models or the derived conditions for optimality into more tractable and independent submodels of lower order. Two reduction techniques are discussed: symmetry reduction and time-scale separation. The symmetry reduction method takes advantage of the geometrical symmetry that may exist in a nuclear reactor configuration. Its foundations are based on the well-developed representation theory of abstract groups. Time-scale separation takes advantage of the large difference between time constraints that is often found in interacting dynamic processes. Its foundation is based on the theory of singular perturbations.

Table 3.9. Inner Product Values

Subsystem		$\langle \psi_{nik}^{(q)}, \phi_0 \rho_j^{(q)} \rangle$		
q	nik	$j = 1$	$j = 2$	$j = 3$
1	101	0.6331×10^{11}	0.1954×10^{14}	0.8030×10^{14}
	121	0.3845×10^{11}	-0.3839×10^{14}	-0.4175×10^{13}
2	12i	0.1686×10^{12}		
	141	0.8859×10^{11}		
3	121	0.1192×10^{12}		
	141	0.6264×10^{11}		
4	101	0.8953×10^{11}	0.2764×10^{14}	0.1136×10^{15}
	121	0.5437×10^{11}	-0.5430×10^{14}	-0.5904×10^{13}
5	111	0.1336×10^{12}		
	131	-0.5708×10^{11}		
6	111	0.6805×10^{11}	0.3554×10^{14}	
	131	0.164×10^{12}	-0.3853×10^{14}	
7	111	0.9624×10^{11}	0.5026×10^{14}	
	131	0.1646×10^{12}	-0.5448×10^{14}	
8	111	0.9448×10^{11}		
	131	-0.4036×10^{11}		

3.4.1. Symmetry Reduction

Geometric symmetry is an inherent property of reactor core designs. In the absence of external disturbances, the power distribution in the core would reach an equilibrium state that exhibits some degree of symmetry. The linearized model, valid for reproducing the core dynamics near the equilibrium state, reflects the same symmetry, and symmetry reduction techniques can drastically simplify the problem of handling the linear model in control applications.

The main benefit of symmetry reduction is the decomposition of the original linear model into a collection of decoupled submodels of lower order. Through symmetry reduction techniques, control systems can be developed for each submodel. Symmetry reduction can be used in conjunction with other model-simplifying procedures, such as modal expansion approximation and Avery's theory of coupled reactors.

Group theory is the standard mathematical tool that treats symmetry systematically. The following section is devoted to the basic concepts of group theory.

3.4.1.1. Some Group-Theoretic Concepts (Ref. 3.27)

This section establishes the operational notation that we will use. Only the basic concepts of group theory that are needed to describe the symmetry reduction technique are cited here.

An *abstract group* is a set of distinct elements with a multiplication law such that

(a) Multiplication is defined for every ordered pair of elements of the group, is closed, single-valued, and associative. That is, the product of two elements of the group is always in the group. If B, C, and D are elements of a group then $BCD = (BC)D = B(CD)$.

(b) There exists an identity element I in the group. That is, there is an element I such that for all other elements A in the group, $IA = AI = A$.

(c) Every element has an inverse relative to the identity I. For each element A in a group there there is an element B in the group such that $AB = BA = I$.

The number of elements in a group is called the *order* of the group. A group is completely defined by specifying the product of every ordered pair of elements. Consider for example the multiplication table

	I	A_1	A_2	A_3
I	I	A_1	A_2	A_3
A_1	A_1	I	A_3	A_2
A_2	A_2	A_3	I	A_1
A_3	A_3	A_2	A_1	I

where the entry in the ith row and jth column is the product of the ith and jth elements of a fourth-order group $G = \{I, A_1, A_2, A_3\}$, which satisfies the properties of an abstract group and is defined by its multiplication table.

If D_1 and D_2 are elements of a group, then the element $D_1^{-1}D_2D_1$ is said to be the *conjugate* of D_2 with respect to D_1.

A *class* of a group is a maximal set of mutually conjugate elements. That is, if $C = \{A_i\}$ is a class of G, then G contains no element that is conjugate to one of the elements A_i of C but is not in C; also, every two elements of C must be conjugate to each other with respect to some element of G.

Consider the eighth-order group Q defined by the multiplication table

	I	A_1	A_2	A_3	A_4	A_5	A_6	A_7
I	I	A_1	A_2	A_3	A_4	A_5	A_6	A_7
A_1	A_1	I	A_3	A_2	A_5	A_4	A_7	A_6
A_2	A_2	A_3	A_1	I	A_6	A_7	A_5	A_4
A_3	A_3	A_2	I	A_1	A_7	A_6	A_4	A_5
A_4	A_4	A_3	A_7	A_6	A_1	I	A_2	A_3
A_5	A_5	A_3	A_6	A_7	I	A_1	A_3	A_2
A_6	A_6	A_3	A_4	A_5	A_3	A_2	A_1	I
A_7	A_7	A_3	A_5	A_4	A_2	A_3	I	A_1

The reader can verify that Q has classes

$$C_1 = I$$
$$C_2 = A_1$$
$$C_3 = \{A_2, A_3\}$$
$$C_4 = \{A_4, A_5\}$$
$$C_5 = \{A_6, A_7\}$$

A *representation* Γ of a group G is a group in which the elements are matrices. The matrices in Γ are associated with the elements of the group G through a many-to-one correspondence that preserves multiplication.

Consider for example the set of matrices $\Gamma = \{I, D_1, D_2, D_3\}$, where

$$I = \begin{bmatrix} 1 & 0 \\ 0 & 1 \end{bmatrix}$$

$$D_1 = \begin{bmatrix} -1 & 0 \\ 0 & 1 \end{bmatrix}$$

$$D_2 = \begin{bmatrix} -1 & 0 \\ 0 & -1 \end{bmatrix}$$

$$D_3 = \begin{bmatrix} 1 & 0 \\ 0 & 1 \end{bmatrix}$$

We can see that the multiplication table for Γ is the same as that for the group G. The group Γ is a representation of G via the correspondence

$$I \to I$$
$$D_1 \to A_1$$
$$D_2 \to A_2$$
$$D_3 \to A_3$$

Two sets of square matrices, say $\{D_i\}$ and $\{D_i'\}$ for $i = 1, 2, \ldots, h$ are equivalent if there exists a square matrix M such that

$$D_i' = M^{-1} D_i M, \qquad i = 1, 2, \ldots, h \tag{3.134a}$$

for some ordering of the matrices $\{D_i'\}$.

A representation of a group is *reducible* if it is equivalent to a representation of G composed of matrices in upper or lower block-triangular form

$$D_i = \begin{bmatrix} D_i^{(11)} & 0 \\ D_i^{(21)} & D_i^{(22)} \end{bmatrix}$$

where the dimension of $D_i^{(11)}$ is equal to the dimension of $D_j^{(11)}$ for every i and j.

Take, for example, the representation Γ; it is clear that Γ is equivalent to itself via any two-dimensional nonsingular matrix M. Since Γ is already composed by matrices in block-diagonal form, it is reducible.

The unidimensional representation $\Gamma_2 = \{-1, -1, 1, 1\}$ of G is an example of an irreducible representation.

The main theorem of the representation of groups states that every group has exactly as many inequivalent irreducible representations as there are classes in the group. Furthermore, if $\Gamma^{(q)} - \{D_j^{(q)}\}$ and $\Gamma^{(p)} = \{D_j^{(p)}\}$, $j = 1, 2, \ldots, h$, are two of these representations, then the matrix entries satisfy the following orthogonality relations:

$$\sum_{j=1}^{h} D_{j_{mn}}^{(q)} [D_j^{(p)}]_{ik}^{-1} = 0 \qquad \text{if } q \neq p \tag{3.134b}$$

$$\text{and/or } n \neq k$$

$$\sum_{j=1}^{h} D_{j_{mn}}^{(q)} [D_j^{(q)}]_{ik}^{-1} = 0 \qquad \text{if } n \neq i \tag{3.135}$$

$$\sum_{j=1}^{h} D_{j_{mn}}^{(q)} [D_j^{(q)}]_{mn}^{-1} = \frac{h}{l_q} \tag{3.136}$$

where $D_{j_{mn}}^{(q)}$ denotes the (mn)th entry in the jth matrix of the representation $\Gamma^{(q)}$, $[D_j^{(q)}]^{-1}$ denotes the inverse of $D_j^{(q)}$, and h and l_q represent the order of the group and the dimension of each of the matrices in $\Gamma^{(q)}$, respectively.

To illustrate, consider the unidimensional matrix group $\Gamma_1 = \{1, 1, 1, 1, -1, -1, -1\}$ and the two-dimensional matrix group $\Gamma_2 = (I, D_1, D_2, D_3, D_4, D_5, D_6, D_7\}$ with

$$I = \begin{bmatrix} 1 & 0 \\ 0 & 1 \end{bmatrix}, \qquad D_1 = \begin{bmatrix} -1 & 0 \\ 0 & -1 \end{bmatrix}, \qquad D_2 = \begin{bmatrix} -j & 0 \\ 0 & j \end{bmatrix}$$

$$D_3 = \begin{bmatrix} j & 0 \\ 0 & -j \end{bmatrix}, \qquad D_4 = \begin{bmatrix} 0 & 1 \\ 0 & j \end{bmatrix}, \qquad D_5 = \begin{bmatrix} 0 & -1 \\ 1 & 0 \end{bmatrix}$$

$$D_6 = \begin{bmatrix} 0 & -j \\ -j & 0 \end{bmatrix}, \qquad D_7 = \begin{bmatrix} 0 & j \\ j & 0 \end{bmatrix}, \qquad j^2 = -1$$

Γ_1 and Γ_2 are both inequivalent, irreducible representations of the example group Q via the correspondence

$$I \to 1, \qquad A_4 \to -1$$
$$A_1 \to 1, \qquad A_5 \to -1$$
$$A_2 \to 1, \qquad A_6 \to -1$$
$$A_3 \to 1, \qquad A_7 \to -1$$

and $D_i \to A_i$.

Let us take Γ_p to be the one-dimensional representation and Γ_q to be the two-dimensional representation. Furthermore, let us choose the (m, n)th entries to be the $(2, i)$th. The first orthogonality relation (3.14a) then becomes

$$1(0) + 1(0) + 1(0) + 1(0) - 1(1) - 1(-1) - 1(-j) - 1(j) = 0$$

Let us choose the (i, k)th and (m, n)th matrix entries to be the $(1, 2)$th and $(2, 2)$th, respectively. The left side of the second orthogonality relation becomes

$$1(0) - 1(0) + j(0) - j(0) + 0(-1) + 0(1) + 0(-j) + 0(j) = 0$$

Similarly, for (m, n) chosen to be $(1, 2)$ the left side of the third orthogonality relation becomes

$$0(0) + (0) + 0(0) + 0(0) + 1(1) - 1(-1) - j(+j) + j(-j) = 4$$

The proof of the theorem is beyond the scope of this book. This group-theoretic result, as abstract as it may seem, forms the foundation of symmetry reduction. The reader should clearly understand the theorem before proceeding.

We conclude this section with a few remarks on representations.

Many of the groups encountered in mathematical physics have been tabulated in the literature. There is, however, no satisfactory method for deriving them in general.

In addition to the matrix elements of the irreducible representation of a group, the traces (characters) satisfy orthogonality relations. To see this, let $n = m$ and $i = k$ in orthogonality relation (3.134a), then sum over i and n to get the traces

$$\sum_{j=1}^{h} (\text{trace } D_j^{(q)})(\text{trace}[D_j^{(p)}]^{-1}) = 0$$

for $q \ne p$. Similarly, from orthogonality relation (3.135) it follows that

$$\sum_{j=1}^{h} (\text{trace } D_j^{(q)})(\text{trace}[D_j^{(q)}]^{-1}) = h$$

3.4.1.2. Symmetry Principles (Refs. 3.22–3.24)

The *symmetry* of a body is described by giving the set of all transformations that preserve distances and bring the body into coincidence with itself. Any such transformation is called a *symmetry transformation.*

For a given body a complete list of symmetry transformations satisfies the group properties. This list is called the *symmetry group* of the body.

If a body is finite in extent, the symmetry transformations can all be built up from three fundamental types:

1. Rotation through a definite angle about some axis (R)
2. Mirror reflection in a plane (σ)
3. Mirror reflection at a center of symmetry (inversion, i)

In what follows, the ordered collection $C = \{C_i\}$ will represent a set of regions, the union of which constitutes a given body. The regions are disjoint, and the only common points between any two contiguous regions are located at their shared boundaries. A symmetry transformation defined on the core, which physically represents a positional interchange of reactor regions such that distances between points of the core are preserved, can be seen to represent in the context of abstract sets of permutation operation defined on the ordered set C.

If R denotes a symmetry transformation on the body, then $RC = \{RC_i\}$ denotes the image of the set C under R, and it is composed of the same collection of regions but now in different order. For example, the rectangle of Figure 3.1, which arbitrarily has been decomposed into four equivalent disjoint regions, has its symmetry described by the following symmetry transformation:

(a) σ_1, reflection with respect to the y-axis

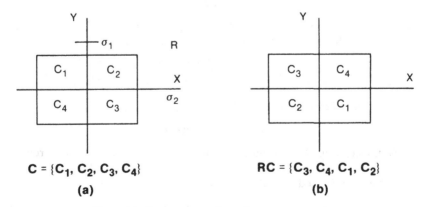

Figure 3.1. Symmetry transformations on a rectangle.

(b) σ_2, reflection with respect to the x-axis
(c) R, rotation of π radians with respect to the center
(d) I, the identity

Applying the transformation R to the rectangle interchanges regions 1 and 3 and regions 4 and 2. Therefore, making use of our notation, it follows that

$$RC_1 = C_3, \quad RC_2 = C_4, \quad RC_3 = C_1, \quad RC_4 = C_2$$

and the ordered set $C = \{C_1, C_2, C_3, C_4\}$ is transformed into $RC = \{C_3, C_4, C_1, C_2\}$.

The dependency of a function $\psi(r)$ on the spatial variable r, defined on the region occupied by the body C, will be explicitly shown in the form $\psi(C)$. This notation is particularly convenient because it facilitates the extension of the concept of a symmetry transformation on a body, or on an ordered set C, to the concept of a symmetry transformation on functions of r. In this manner, if R is a symmetry transformation belonging to the symmetry group of a body C, then, by definition, the image $R\psi(C)$ of a function $\psi(C)$ under R is given by the function ψ defined over the transformed body, RC. This suggests the notation

$$R\psi(C) \triangleq RC$$

If $\psi(C)$ is an element of some linear space, then it is clear that R defines a linear transformation from the same linear space into itself.

A function defined on C could be defined on a continuum, as for the neutron flux distribution, or as discrete locations, as for the reactor control rods. For example, the distribution $\psi(C)$ defined on the rectangle of Figure 3.1 by

$$\psi(x, y) = \cos\left(\frac{\pi x}{a}\right) \sin\left(\frac{2\pi y}{b}\right) \tag{3.137}$$

where x and y take on values over $[-a/2, a/2]$ and $[-b/2, b/2]$, respectively, is transformed by the reflection operation σ_2 into

$$\sigma_2(x, y) = -\cos\left(\frac{\pi x}{a}\right) \sin\left(\frac{2\pi y}{b}\right) \tag{3.138}$$

which, in our compact notation, becomes

$$\sigma_2\psi(C) = -\psi(C) \tag{3.139}$$

Similarly, if u_i denotes some specific property associated with the region C_i in the rectangle of Figure 3.1, then the vector $U(C) = \text{col}[u_1, u_2, u_3, u_4]$ would be transformed by σ_2 as follows:

$$\sigma_2 U(C) = \text{col}[u_4, u_3, u_2, u_1] \tag{3.140}$$

A collection of functions $\{f_j(C)\}$, finite or infinite, is said to generate a representation $\{D_i\}$ of a symmetry group $g = \{R_i\}$ associated with body C, if for every symmetry transformation R_i and every function $f_j(C)$ of the term $R_i f_j(C)$ satisfies a relation of the form

$$R_i f_j(C) = \sum_k D_{i_{jk}} \cdot f_k(C) \tag{3.141}$$

via the correspondence $D_i \to R_i$. To prove that $\{D_i\}$ is indeed a representation, we need to show that it satisfies the group properties. This can be done by means of a column vector F with entries $f_i(C)$. We prove it, for example, for closure of $\{D_i\}$ under matrix multiplication.

Let R_k and R_l be any two symmmetry transformations in $\{R_i\}$, then from the conditions

$$R_k F = D_k F$$

$$R_l F = D_l F$$

$$R_k R_l F = D_k D_l F$$

and since the product $R_k R_l$ is in $\{R_i\}$, it follows that $D_k D_l$ is in $\{D_i\}$, proving that matrix multiplication is closed. Similarly, the existence in $\{D_i\}$ of the identity and the inverse matrices D_i^{-1} follows from the group properties of $\{R_i\}$.

Any function $f(C)$ belonging to a linear space H defined on a body C having a symmetry group $G = \{R_i\}$ can generate a representation of G. To see this, let f_i denote the function $f(R_i C)$. Then the set $\{f_i\}$, $i = 1, 2, \ldots, h$, defines a collection of functions that also belong to H. Since, for any transformation R_i in the group G and any function f_j the product $R_i f_j$ is also in the set $\{f_i\}$, it follows that the project $R_i f_j$ can be represented in the form

$$R_j f_i = \sum_{k=1}^{l} D_{jik} f_k \tag{3.142}$$

where l is the number of unrepeated functions in the set $\{f_i\}$, and D_j denotes the jth matrix in the representation of G generated by the function F.

To illustrate, consider the function $q(C)$ defined on the rectangle of Figure 3.1, where

$$q = z + w$$

and

$$a(x, y) = \cos\left(\frac{\pi x}{a}\right) \sin\left(\frac{2\pi y}{b}\right)$$

$$w(x, y) = \sin\left(\frac{2\pi y}{a}\right)$$

Via the correspondence $\{q_i\} \to \{I_q, \sigma, q, \sigma_2 q, rq\}$,

$$q_1 = z + w$$
$$q_2 = z - w$$
$$q_3 = -z + w$$
$$q_4 = -z - w$$

we find that $\{q_i\}$ generates a representation of $\{I, \sigma_1, \sigma_2, R\}$. For example,

$$\sigma_1 q_2 = (1)y_1 + (0)y_2 + (0)y_3 + (0)y_4$$

In general, a representation of a given symmetry group, generated by an arbitrary function, is reducible. We are now ready to profit from the main orthogonality theorem of representation theory.

We recall that a linear space H is said to be decomposed into a direct sum of subspaces H, if each subspace is disjoint from the subspaces spanned by the others and the union of all subspaces is H. Furthermore, if f is any element in H, then f can be represented uniquely by the sum of components

$$f = \sum_i f_i$$

where f_i belongs to H_i.

Suppose now that H is a linear space of functions defined on a body C, having symmetry group $G = \{R_i\}$, order h, and N_C classes (therefore the same number of irreducible, inequivalent representation). Let l_q denote the dimension of the matrices in the qth irreducible representation $\Gamma^{(q)}$. Then, by a direct application of the main representation theorem, it follows that H can be decomposed into a direct sum of subspaces $H_n^{(q)}$ of functions that generate the nth row of the qth irreducible representation (q).

In other words, any arbitrary function $f(C)$ in H can be uniquely represented by

$$f(C) = \sum_{n=1}^{l_q} \sum_{q=1}^{N_C} f_n^{(q)}(C) \tag{3.143}$$

according to the form

$$R_i f_n^{(q)}(C) = \sum_{k=1}^{l_q} \Gamma_{ink}^{(q)} f_k^{(q)}(C) \tag{3.144}$$

in which $\Gamma_{ink}^{(q)}$ denotes the nkth entry in the ith matrix of the qth irreducible representation of G.

To prove this statement, it suffices to give the collection of projection operators

$$P_n^{(q)} = \frac{l_q}{h} \sum_{i=1}^{h} [\Gamma_{inn}^{(q)}]^{-1} R_i \tag{3.145}$$

where $[\Gamma_{inn}^{(q)}]^{-1}$ denotes the nth diagonal entry in the inverse of $\Gamma_i^{(q)}$.

Consider, for example, an arbitrary function f in H with component $f_n^{(q)}$ in $H_n^{(q)}$ and represented by

$$f = \sum_{n=1}^{l_q} \sum_{q=1}^{N_C} f_n^{(q)} \tag{3.146}$$

Then

$$P_m^{(p)} f = \frac{l_p}{h} \sum_{i=1}^{h} [\Gamma_{i_{mm}}^{(q)}]^{-1} \sum_{n=1}^{l_q} \sum_{q=1}^{N_C} F_i f_n^{(q)} \tag{3.147}$$

After substituting (3.144) in (3.147) and reordering, we find

$$P_m^{(p)} f = \frac{l_p}{h} \sum_{n=1}^{l_q} \sum_{q=1}^{N_C} \sum_{k=1}^{l_q} \sum_{i=1}^{h} [\Gamma_{i_{mm}}^{(p)}]^{-1} \Gamma_{i_{nk}}^{(q)} f_k^{(q)} \tag{3.148}$$

and making use of the orthogonality relations (3.134) to (3.135) of the main theorem, it follows that

$$P_m^{(p)} f = \frac{l_p}{h} \sum_{i=1}^{h} [\Gamma_{i_{mm}}^{(p)}]^{-1} \Gamma_{i_{mm}}^{(p)} f_m^{(p)}$$

$$= \frac{l_p}{h} \frac{h}{l_p} f_m^{(p)}$$

$$= f_m^{(p)} \tag{3.149}$$

This shows that $H_m^{(p)}$ is indeed the range of $P_m^{(p)}$. In a similar manner, we can show that

$$P_n^{(q)} P_m^{(p)} = 0 \qquad \text{if } n \neq m \text{ or } p \neq q \tag{3.150}$$

$$P_n^{(q)} P_n^{(q)} = P_n^{(q)} \tag{3.151}$$

and

$$\sum_{q=1}^{N_C} \sum_{n=1}^{l_q} P_n^{(q)} = \text{identity} \tag{3.152}$$

therefore proving that the subspaces $\{H_n^{(q)}\}$ decompose H in a direct sum manner.

We shall end this section with some general remarks. Recall that disjoint subspaces of an inner-product space are orthogonal; i.e., the inner product between elements of disjoint subspaces is zero. Therefore if our linear space H is endowed with an inner product, then the subspaces $H_n^{(q)}$ would not only be disjoint but orthogonal.

We emphasize that the decomposition of a linear space H of functions defined on a body C having symmetry group G is derived from properties of the irreducible representation of G. Again, many of the groups encountered in mathematical physics have been tabulated in the literature.

Chemists have developed two notational systems (Schönflies and International) to classify symmetry groups into a few types. It is in terms of these notation systems that the irreducible representations of symmetry groups can most easily be found in the literature.

3.4.1.3. Symmetry Reduction of Nuclear Reactor Models (Refs. 3.28–3.32)

The application of the group-theoretic results of the previous section can drastically simplify the analysis and treatment of those problems involving linear distributed parameter models that, due to the nature of the physical system that they represent, exhibit some degree of symmetry. Consider, for example, the general linearized distributed reactor model

$$\frac{\partial \psi}{\partial t}(r, t) = A(r)\psi(r, t) + B(r)U(t) \tag{3.153}$$

where r is the spatial variable defined on the reactor core C, and ψ denotes the state of the reactor core. For fixed r and t, $\psi(r, t)$ represents an N-dimensional vector, and $A(r)$ is a matrix spatial operator involving operators of the diffusion type. For fixed time t, $U(t)$ is an M-dimensional vector representing the effect of the control rods, and $B(r)$ is a rectangular matrix spatial operator of the appropriate dimension. Also associated with equation (3.153) are boundary conditions

$$\psi(r, t) = 0 \tag{3.154}$$

at the boundary of the core, and the initial condition

$$\psi(r, t_0) = Z_0(r) \tag{3.155}$$

at time t_0.

In our new notation equations (3.153) to (3.155) are

$$\frac{\partial}{\partial t}\psi(c, t) = A(c)\psi(c, t) + B(c)U(c, t) \tag{3.156}$$

$$\psi(c, t) = 0 \qquad \text{at the boundary} \tag{3.157}$$

and

$$\psi(c, t_0) = Z_0(c) \tag{3.158}$$

Although the vector U does not depend explicitly on the spatial variable r, it clearly represents a function of the core, since every entry in U corresponds to a control rod, which in turn is associated with a certain location in the core.

If the geometrical configuration of the reactor has symmetry group $G = \{R_i\}$, and if the reactor parameters in the mathematical model are distributed over the core according to the same symmetry described by G, then it follows that the operators A and B are invariant under symmetry transformations R_i.

Since A and B are linear, it is clear that the application of the projection operators $\{p_n^{(q)}\}$ to equations (3.156) to (3.158) yields

$$\frac{\partial}{\partial t} \psi_n^{(q)}(c, t) = A(c)\psi_n^{(q)}(c, t) + B(c)U_n^{(q)}(c, t) \tag{3.159}$$

$$\psi_n^{(q)}(c, t) = 0 \qquad \text{at the boundary} \tag{3.160}$$

and

$$\psi_n^{(q)}(c, t) = Z_{0n}^{(q)}(c) \tag{3.161}$$

where

$$\psi(c, t) = \sum_{q=1}^{N_C} \sum_{n=1}^{l_q} \psi_n^{(q)}(c, t) \tag{3.162}$$

and

$$U(c, t) = \sum_{q=1}^{N_C} \sum_{n=1}^{l_q} U_n^{(q)}(c, t) \tag{3.163}$$

Thus, the projection operations $\{p_n^{(q)}\}$ decompose the space H of state distributions and the space E of control vectors into the invariant manifolds $H_n^{(q)}$ and $E_n^{(q)}$ of $A(c)$ and $B(c)$, respectively. It follows from this observation that optimal control problems involving a performance index of this type are given by

$$J(U) = \int_{t_0}^{t_1} [\|\psi - Z\|_H + k_0\|U\|_E] \, dt \tag{3.164}$$

and mathematical models such as equation (3.153) can be reduced to a number of subproblems, which consist of minimizing

$$J_n^{(q)}(U) = \int_{t_0}^{t_1} [\|\psi - Z\|_{H_n^{(q)}} + k_0\|U\|_{E_n^{(q)}}] \, dt \tag{3.165}$$

subject to (3.159), (3.160), and (3.161).

The benefits of the reductive approach are clear. Although the original problem implies a search for M functions of time (the entries $u_i(t)$ of the control vector U) in the case of the subproblems, it is the dimension of the subspaces $E_n^{(q)}$ that indicates the number of functions of time that ought to be computed. Depending on the degree of symmetry exhibited by the reactor core, these subspaces could have a dimension substantially lower than M.

3.5. An Example: Decoupling of the Optimality Conditions (Refs. 3.25–3.28)

We illustrate the benefits of using symmetry principles to treat optimality conditions that result from applying optimization techniques, such as the minimum norm formulation.

Recall that the necessary and sufficient conditions for optimality that give the solution to the problem of minimizing the performance index

$$J(U) = \int_{t_0}^{t_1} \int_0^b \psi^2(r, t) \, dr \, dt + k_0 \int_{t_0}^{t_1} \sum_{i=1}^M u_i^2(t) \, dt \qquad (3.166)$$

subject to the slab reactor model

$$\frac{\partial \psi(r, t)}{\partial t} = VD \frac{\partial^2}{\partial r^2} \psi(r, t) + V[v\Sigma_f - \Sigma_a]\psi(r, t)$$

$$- V \sum_{i=1}^M u_i(t)\phi_0(r)\delta(r - r_i) \qquad (3.167)$$

$$\phi_0(r) = \phi_M \sqrt{\frac{2}{b}} \sin\left(\frac{\pi r}{b}\right) \qquad (3.168)$$

with boundary conditions

$$\psi(0, t_0) = \psi(b, t) = 0 \qquad (3.169)$$

and initial condition

$$\psi(r, t_0) = Z_0(r) \qquad (3.170)$$

are given by the coupled set of Fredholm integral equations

$$u_i(\tau) = \Delta_i(\tau) - k_0^{-1} \int_{t_0}^{t_1} \sum_{j=1}^m \hat{K}_{ij}(\tau; \alpha)u_j(\alpha) \, d\alpha \qquad i = 1, 2, \ldots, M \qquad (3.171)$$

with kernels defined by

$$\hat{K}_{ij}(\tau, \alpha) = \frac{\phi_M^2 4V^2}{b^2} \sum_{n=1}^\infty \sin\left(\frac{n\pi r_i}{b}\right) \sin\left(\frac{\pi r_i}{b}\right) \sin\left(\frac{\pi r_j}{b}\right) \sin\left(\frac{n\pi r_j}{b}\right) K_n(\tau; \alpha) \qquad (3.172a)$$

$$K_n(\tau; \alpha) = \begin{cases} [\exp \lambda_n(2t_1 - \tau - \alpha) - \exp \lambda_n|\tau - \alpha|]/2\lambda_n & \text{for } \lambda_n \neq 0 \\ t_1 - \tau & \text{for } \lambda_n = 0 \text{ and } \alpha < \tau \\ t_1 - \alpha & \text{for } \lambda_n = 0 \text{ and } \alpha \geq \tau \end{cases}$$

(3.172b)

The forcing functions $\Delta_i(\tau)$ are given by

$$\Delta_i(\tau) = \sum_{n=1}^{\infty} f_n(\tau) \sin\left(\frac{n\pi r_i}{b}\right) \sin\left(\frac{\pi r_i}{b}\right) \int_0^b \sin\left(\frac{n\pi r'}{b}\right) Z_0(r') \, dr' \quad (3.173)$$

where

$$f_n(\tau) = k_0^{-1} \phi_M 2V[\exp \lambda_n(\tau - t_0) - \exp \lambda_n(2t - t_0 - \tau)]\lambda_n^{-1} \quad (3.174)$$

and

$$\lambda_n = \left[v\Sigma_f - \Sigma_a - \frac{n^2\pi^2}{b^2} D \right] V \quad (3.175)$$

If the control planes $r = r_i$, $i = 1, 2, \ldots, M$, are placed symmetrically with respect to the center plane of the reactor, $r = b/2$, one finds that the core would have a symmetry group G composed of two symmetry transformations: the identity I and the mirror reflection with respect to the center plane, σ. That $\{I, \sigma\}$ satisfies the properties of a group is clear from the relations

$$II = I \quad (3.176)$$

$$I\sigma = \sigma I = \sigma \quad (3.177)$$

and

$$\sigma\sigma = I \quad (3.178)$$

This group is identified in mathematical physics as the symmetry group C_s (Ref. 3.15).

The group G has the two unidimensional irreducible representations

	I	σ
Γ_1	1	1
Γ_2	1	-1

The projection operators $P_n^{(q)}$ are

$$P_1^{(1)} = \tfrac{1}{2}[I + \sigma] \quad (3.179)$$

and

$$P_1^{(2)} = \tfrac{1}{2}[I - \sigma] \quad (3.180)$$

For simplicity, suppose there are only two control rods (planes) symmetrically located with respect to the center plane. Then it follows that the kernels $\hat{K}_{ij}$ in the optimality conditions (3.171) satisfy the relations

$$\hat{K}_{12}(\tau; \alpha) = \hat{K}_{21}(\tau; \alpha) \tag{3.181}$$

and

$$\hat{K}_{11}(\tau; \alpha) = \hat{K}_{22}(\tau; \alpha) \tag{3.182}$$

Also, since the effect of applying the transformation σ on the reactor is felt upon the optimality conditions as the interchange of subscripts 1 and 2, it follows that $\hat{K}_{ij}$ are invariant under σ. Therefore, it is possible to reduce the optimality conditions into two decoupled equations by symmetry considerations.

With only two control functions, $u_1(t)$ and $u_2(t)$, the space E becomes the two-dimensional Euclidean space. Applying the projections $p_n^{(q)}$ to any base in E, one finds that the invariant manifolds $E_1^{(1)}$ and $E_1^{(2)}$ are the unidimensional subspaces spanned by the normalized vectors

$$\rho_1^{(1)} = \frac{1}{\sqrt{2}}\begin{bmatrix} 1 \\ 1 \end{bmatrix} \quad \text{and} \quad \rho_1^{(2)} = \frac{1}{\sqrt{2}}\begin{bmatrix} 1 \\ 1 \end{bmatrix}$$

respectively. Thus, by expanding the control vector U in the form

$$U(t) = \sum_{q=1}^{2} \beta_1^{(q)}(t)\rho_1^{(q)} \tag{3.183}$$

where the $\{\beta_1^{(q)}\}$ are the time-dependent expansion coefficients, and substituting (3.183) in the optimality conditions, one finds, after taking the inner products in E, that the conditions for optimality become

$$\beta_1^{(1)}(\tau) = \frac{1}{\sqrt{2}}[\Delta_1(\tau) + \Delta_2(\tau)] - k_0^{-1} \int_{t_0}^{t_1} [\hat{K}_{11}(\tau, \alpha) + \hat{K}_{21}(\tau; a)]\beta_1^{(1)}(\alpha)\, d\alpha \tag{3.184}$$

and

$$\beta_1^{(2)}(\tau) = \frac{1}{\sqrt{2}}[\Delta_1(\tau) - \Delta_2(\tau)] - k_0^{-1} \int_{k_0}^{t_1} [\hat{K}_{11}(\tau; a) - \hat{K}_{21}(\tau; \alpha)]\beta^{(2)}(\alpha)\, d\alpha \tag{3.185}$$

These equations can be solved independently. Given the simplicity of our example, the decoupling of the optimality conditions could perhaps have been achieved by other techniques.

However the reduction of mathematical models for reactors with 2- or 3-dimensional configurations becomes more difficult, and the benefits of a systematic approach are welcome. The next sample illustrates the application of symmetry principles to a three-dimensional reactor model.

3.6. An Example: Reduction of a Cylindrical Reactor Model (Ref. 3.29)

Consider the homogeneous cylindrical reactor of Figure 3.2 and the one-energy model

$$\frac{\partial \psi}{\partial t}(r, \theta, z, t) = VD\nabla^2\psi(r, \theta, z, t) + V[v\Sigma_f - \Sigma_a]\psi(r, \theta, z, t) - B(r, \theta, z)U(t) \tag{3.186}$$

where

$$B(r, \theta, z)U(t) = V\phi_0(r, \theta, z) \sum_{i=1}^{14} \delta(z - z_i)\alpha_i(r, \theta)u_i(t) \tag{3.187}$$

and

$$\phi_0(r, \theta, z) = \phi_M \cos\left(\frac{\pi}{2} z\right) J_0(\gamma_{01}r) \tag{3.188}$$

The function $\alpha_i(r, \theta)$ is equal to 1 in the small region occupied by the ith neutron control device, and equal to 0 elsewhere. The operator ∇^2 is the Laplacian in cylindrical coordinates, L is the length of the core, γ_{01} represents the first root of the zero-order Bessel function $J_0(\gamma R)$, and R is the radius of the cylinder. All the other variables have their usual meanings. The flux is equal to zero at the boundary of the core, and the initial condition

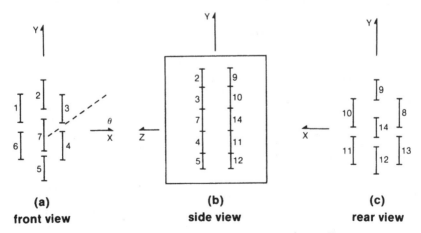

Figure 3.2. Cylindrical reactor core and layout of controllers.

is

$$\psi(r, \theta, z, t_0) = Z_0(r, \theta, z) \tag{3.189}$$

Given the symmetrical allocation of the control devices in the core, as shown in Figure 3.2, one may conclude that the symmetry principles discussed in the previous sections can be applied to the problem of reducing the mathematical model. The reactor core has the following group of symmetry operations: the identity I, counterclockwise rotation of π radians with respect to the x-, y-, and z-axes, denoted, respectively, by R_x, R_y, and R_z. Reflection with respect to the xy-, yz-, and zx-planes is denoted respectively by F_{xy}, F_{yz}, and F_{zx}. The inversion operation with respect to the center is denoted by R_I.

These symmetry transformations, together with the group multiplication table shown in Table 3.10, constitute the symmetry group D_{2h} of mathematical physics. Each symmetry operation forms a class in itself. The eight unidimensional irreducible representations of D_{2h} are given in Table 3.11.

The mathematical model of the core can be reduced into eight subsystems by the projection operators $p_1^{(q)}$, $q = 1, 2, \ldots, 8$. The space E of control vectors becomes the 14-dimensional Euclidean space. The vector $U = \text{col}[u_1, \ldots, u_{14}]$ transforms under the symmetry operations as shown in Table 3.12. For simplicity only the subscript indices are shown. Table 3.12 shows for example that the 6th and 11th entries in U are interchanged by the operation R_y.

By applying the projection operators $p_1^{(q)}$ to any complete set of vectors in E, one finds that the eight invariant subspaces $\{E_1^{(q)}\}$, $q = 1, 2, \ldots, 8$, are spanned by the vectors $\{\rho_{1_i}^{(q)}\}$ shown in Table 3.13. Also, the eigenfunctions of the spatial operator

$$VD\nabla^2 + V[v\Sigma_f - \Sigma_a] \tag{3.190}$$

Table 3.10. Group Multiplication Table. Sample: $R_I R_x = F_{yz}$

D_{2h}	I	R_I	R_x	R_y	R_z	F_{zx}	F_{xy}	F_{yz}
I	I	R_I	R_x	R_y	R_z	F_{zx}	F_{xy}	F_{yz}
R_I	R_I	I	F_{yz}	F_{zx}	F_{xy}	R_y	R_z	R_x
R_x	R_x	F_{yz}	I	R_z	R_y	F_{xy}	F_{zx}	R_I
R_y	R_y	F_{zx}	R_z	I	R_x	R_I	F_{yz}	F_{xy}
R_z	R_z	F_{xy}	R_y	R_x	I	F_{yz}	R_I	F_{zx}
F_{zx}	F_{zx}	R_y	F_{xy}	R_I	F_{yz}	I	R_x	R_z
F_{xy}	F_{xy}	R_z	F_{zx}	F_{yz}	R_I	R_x	I	R_y
F_{yz}	F_{yz}	R_x	I	F_{xy}	F_{zx}	R_z	R_y	I

Table 3.11. Irreducible Representations of D_{2h}

D_{2h}	I	R_z	R_y	R_x	R_I	F_{xy}	F_{xz}	F_{yz}
$\Gamma^{(1)}$	1	1	1	1	1	1	1	1
$\Gamma^{(2)}$	1	1	1	1	-1	-1	-1	-1
$\Gamma^{(3)}$	1	1	-1	-1	1	1	-1	-1
$\Gamma^{(4)}$	1	1	-1	-1	-1	-1	1	1
$\Gamma^{(5)}$	1	-1	1	-1	1	-1	1	-1
$\Gamma^{(6)}$	1	-1	1	-1	-1	1	-1	1
$\Gamma^{(7)}$	1	-1	-1	1	1	-1	-1	1
$\Gamma^{(8)}$	1	-1	-1	1	-1	1	1	-1

Table 3.12. Transformation of the Vector U. Sample; $R_y u_6 = u_{11}$

	u_1	u_2	u_3	u_4	u_5	u_6	u_7	u_8	u_9	u_{10}	u_{11}	u_{12}	u_{13}	u_{14}
I	1	2	3	4	5	6	7	8	9	10	11	12	13	14
R_z	4	5	6	1	2	3	4	11	12	13	8	9	10	14
R_y	10	9	8	13	12	11	14	3	2	1	6	5	4	7
R_x	13	12	11	10	9	8	14	6	5	4	3	2	1	7
R_I	11	12	13	8	9	10	14	4	5	6	1	2	3	7
F_{xy}	8	9	10	11	12	13	14	1	2	3	4	5	6	7
F_{zx}	6	5	4	3	2	1	7	13	12	11	10	9	8	14
F_{yz}	3	2	1	6	5	4	7	10	9	8	13	12	11	14

Table 3.13. The Invariant Subspaces $E_1^{(q)}$ and the Base Vectors $\rho_{1m}^{(q)}$

$E_1^{(1)}$			$E_1^{(2)}$	$E_1^{(3)}$	$E_1^{(4)}$			$E_1^{(5)}$	$E_1^{(6)}$		$E_1^{(7)}$		$E_1^{(8)}$
$\rho_{11}^{(1)}$	$\rho_{12}^{(1)}$	$\rho_{13}^{(1)}$	$\rho_{11}^{(2)}$	$\rho_{11}^{(3)}$	$\rho_{11}^{(4)}$	$\rho_{12}^{(4)}$	$\rho_{13}^{(4)}$	$\rho_{11}^{(5)}$	$\rho_{11}^{(6)}$	$\rho_{12}^{(6)}$	$\rho_{11}^{(7)}$	$\rho_{12}^{(7)}$	$\rho_{11}^{(8)}$
1	0	0	1	1	1	0	0	1	1	0	1	0	1
0	1	0	0	0	0	1	0	0	0	0	0	1	0
1	0	0	-1	-1	1	0	0	-1	1	1	1	0	-1
1	0	0	1	1	1	0	0	-1	-1	-1	-1	0	-1
0	1	0	0	0	0	1	0	0	0	0	0	-1	0
1	0	0	-1	-1	1	0	0	1	-1	-1	-1	0	1
0	0	1	0	0	0	0	1	0	0	0	0	0	0
1	0	0	-1	1	-1	0	0	-1	1	-1	-1	0	1
0	1	0	0	0	0	-1	0	0	0	0	0	-1	0
1	0	0	1	-1	-1	0	0	1	1	-1	-1	0	-1
1	0	0	-1	1	-1	0	0	1	-1	1	1	0	-1
0	1	0	0	0	0	-1	0	0	0	0	0	1	0
1	0	0	1	-1	-1	0	0	-1	-1	1	1	0	1
0	0	1	0	0	0	0	-1	0	0	0	0	0	0

generate all the irreducible representations of the group D_{2h}. Table 3.14 shows the classification of the eigenfunctions according to the irreducible representation that they generate.

The set of eigenfunctions is orthogonal in the real Hilbert space H endowed with the inner product

$$\langle \psi, \phi \rangle = \int_{-1/2}^{L/2} \int_0^{2\pi} \int_0^z r\psi(r, \theta z, z)\phi(r, \theta, z) \, dr \, d\theta \, dz \qquad (3.191)$$

Using eigenfunction expansion, we find that the component $\psi_1^{(q)}(r, \theta, z, t)$ of the flux $\psi(r, \theta, z, t)$, which lies in the invariant subspace $H_1^{(q)}$ spanned

Table 3.14. Laplacian Eigenfunctions $\psi_{nik}^{(q)}$

Subsystem q	Laplacian Modes $\psi_{nik}^{(q)}(r, \theta, z)$			
1	$f_{ik} \cos\left((2n-1)\dfrac{\pi}{L} z\right)$	$\cos(i\theta)$	$J_i(\gamma_{ik} r)$	$i = 0, 2, 4, \ldots$
2	$f_{ik} \sin\left(2n\dfrac{\pi}{L} z\right)$	$\sin(i\theta)$	$J_i(\gamma_{ik} r)$	$i = 2, 4, 6, \ldots$
3	$f_{ik} \cos\left((2n-1)\dfrac{\pi}{L} z\right)$	$\sin(i\theta)$	$J_i(\gamma_{ik} r)$	$i = 2, 4, 6, \ldots$
4	$f_{ik} \sin\left(2n\dfrac{\pi}{L} z\right)$	$\cos(i\theta)$	$J_i(\gamma_{ik} r)$	$i = 0, 2, 4, \ldots$
5	$f_{ik} \sin\left(2n\dfrac{\pi}{L} z\right)$	$\cos(i\theta)$	$J_i(\gamma_{ik} r)$	$i = 1, 3, 5, \ldots$
6	$f_{ik} \cos\left((2n-1)\dfrac{\pi}{L} z\right)$	$\sin(i\theta)$	$J_i(\gamma_{ik} r)$	$i = 1, 3, 5, \ldots$
7	$f_{ik} \sin\left(2n\dfrac{\pi}{L} z\right)$	$\sin(i\theta)$	$J_i(\gamma_{ik} r)$	$i = 1, 3, 5, \ldots$
8	$f_{ik} \cos\left((2n-1)\dfrac{\pi}{L} z\right)$	$\cos(i\theta)$	$J_i(\gamma_{ik} r)$	$i = 1, 3, 5, \ldots$

where

$$n = 1, 2, 3, \ldots \qquad k = 1, 2, 3, \ldots$$
$$-L/2 \le z \le L/2 \qquad 0 \le \theta \le 2\pi$$
$$\gamma_{ik} \text{ is the } k\text{th root of } J_i(\gamma R)$$
$$f_{ik} = \begin{cases} 2/\sqrt{\pi L} \; R J_{i+1}(\gamma_{ik} R) & \text{if } i \neq 0 \\ \sqrt{2}/\sqrt{\pi L} \; R J_{i+1}(\gamma_{ik} R) & \text{if } i = 0 \end{cases}$$

by the infinite set $\{\psi_{nik}^{(q)}\}$, is

$$\psi_1^{(q)}(r, \theta, z, t) = \sum_{(nik)\in\Omega^{(q)}} \psi_{nik}^{(q)}(r, \theta, z)\left[\langle Z_0, \psi_{nik}^{(q)}\rangle H \exp\lambda_{nik}(t - t_0)\right.$$

$$\left. + \sum_{m=1}^{N_q} \int_{t_0}^{t} \langle\beta\rho_{1m}^{(1)}, \psi_{nik}^{(q)}\rangle H \exp\lambda_{nik}(t - \tau)\beta_{1m}^{(q)}(\tau)\,d\tau\right] \quad (3.192)$$

where $\Omega^{(q)}$ denotes the infinite set of subscripts associated with the eigenfunctions in $H_1^{(q)}$, $\lambda_{nik}^{(q)}$ denotes the eigenvalue associated with $\psi_{nik}^{(q)}$,

$$\lambda_{nik}^{(q)} = -VD[\gamma_{ik}^2 + \alpha_{nik}^{(q)^2}] + [v\Sigma_f - \Sigma_a]V \quad (3.193)$$

and

$$\alpha_{nik}^{(q)} = \begin{cases} (2n - 1)\dfrac{\pi}{L} & \text{for } q = 1, 2, 6, 8 \quad (3.194) \\[2mm] \dfrac{2n\pi}{L}, & \text{for } q = 2, 4, 5, 7 \quad (3.195) \end{cases}$$

Also, $\beta_1^{(q)}$ denotes the expansion coefficient corresponding to the base vector $\rho_{1m}^{(q)}$ in the expansion of the control U,

$$U(\tau) = \sum_{q=1}^{8} \sum_{m=1}^{N_q} \beta_{1m}^{(q)}(\tau)\rho_{1m}^{(q)} \quad (3.196)$$

and N_q is the dimension of the invariant subspace $E_1^{(q)}$. Although the dimension of the control space E in the original model is 14, the dimension of the largest subspaces, $E_1^{(1)}$ and $E_1^{(4)}$, is only 3.

3.7. Time-Scale Separation (Ref. 3.29)

Realistic dynamic models as required in many applications exhibit many dynamic patterns of behavior. However, frequently only those near the extremes (fastest or slowest) can be termed relevant for the intended application. This is so, for instance, with the xenon control problem; the relevant dynamic pattern concerns power oscillations, and our interest is limited to the slow, poorly damped natural modes.

The basic idea of time-scale separation is to develop a reduced-order model to reproduce satisfactorily the relevant dynamics of the original model. Singular perturbation techniques can be used to this purpose when the rate of change of some state variables is conditioned on a small parameter (Refs. 3.25 and 3.26).

References

3.1. AVERY, R., "Theory of Coupled Reactors," *Proceedings of the International Conference on the Peaceful Uses of Atomic Energy*, Geneva, **12**, pp. 182-190 (1958).

3.2. ALDER, F. T., GAGE, S. J., and HOPKINS, G. C., "Spatial and Spectral Coupling Effects in Multicore Reactor Systems," *Proceedings of the Conference on Coupled Reactor Kinetics*, edited by C. G. Chezem and W. H. Kohler, Texas A & M Press, College Station, Texas, 1967.

3.3. CHIANG, A. T., BENNECKE, P. E., and RYDIN, R. A., "Optimal Control of Xenon Spatial Oscillations in Reactors," *Transactions of the American Nuclear Society* **24**, 427-433 (1976).

3.4. EL-BASIONI, A. A., and PONCELET, C. G., "Minimal Time Control of Xenon Spatial Oscillations on Reactors," *Nuc. Sci. Eng.* **54**, 166-171 (1974).

3.5. GREEN, R. E., ARGUE, D. S., and LAWRENCE, C. B., "*Spatial Reactor Models for Nuclear Power Plant Simulations*," IAEA Specialist Meeting on Spatial Control Problems, Studvisk, Sweden, 1975.

3.6. HANSEN, K. F., and KANG, C. M., "Finite Element Methods in Reactor Physics Analysis," *Adv. Nucl. Sci. Tech.* **3**, 233-240 (1965).

3.7. IWAZUMI, T., and KOGA, R., "Optimal Feedback Control of a Nuclear Reactor as a Distributed Parameter System," *J. Nucl. Sci. Tech.* **10**(11), 674-680 (1973).

3.8. KAPLAN, S., "The Property of Finality and the Analysis of Problems in Reactor Space-Time Kinetics by Various Modal Expansions," *Nucl. Sci. Eng.* **9**, 357 (1961).

3.9. KAPLAN, S., "Synthesis Methods in Reactor Analysis", *Adv. Nucl. Sci. Tech.* **3**, 233-241 (1966).

3.10. NIEVA, R., CHRISTENSEN, G. S., and EL-HAWARY, M. E., "Suboptimal Control of a Nuclear Reactor Using Functional Analysis," *Int. J. Control* **36**(1), 145-152 (1977).

3.11. ONEGA, R. J., and KISNER, R. A., "A Two-Point Xenon Oscillation Model Using a Variational Principle," *Trans. Am. Nucl. Soc.* **24**, 431-440 (1976).

3.12. STACEY, W. M., "A General Model Expansion Method for Obtaining Approximate Equations for Linear Systems," *Nucl. Sci. Eng.* **28**, 438-443 (1967).

3.13. STACEY, W. M., *Space-Time Nuclear Reactor Kinetics*, Academic Press, New York, 1969.

3.14. STACEY, W. M., "Application of Variational Synthesis to the Optimal Control of Spatially Dependent Reactor Models," *Nucl. Sci. Eng.* **39**, 226-233 (1970).

3.15. TSOURI, N., ROOTENBERGE, J., and LIDOFSKY, L. J., "Optimal Control of a Large Core Reactor in Presence of Xenon," *IEEE Transactions on Nuclear Science* **NS-22**, 702-715 (1975).

3.16. VARGA, R. S., and MARTINO, M. A., "The Theory for the Numerical Solution of Time-Dependent and Time-Independent Multigroup Diffusion Equation," *Proceedings of the International Conference on Peaceful Uses of Atomic Energy*, Geneva, **16**, pp. 570-578 (1958).

3.17. WACHPRESS, E. L., *Iterative Solution of Elliptic Systems*, Prentice-Hall, Englewood Cliffs, New Jersey, 1966.

3.18. WIBERG, D. M.,"Optimal Control of Nuclear Reactor Systems," *Adv. in Control Syst. Theory and Applications* **5**, 301-388 (1967).

3.19. ASATANI, K., SHIOTANI, M., and HATTORI, Y., "Suboptimal Control of Nuclear Reactors with Distributed Parameters Using Singular Perturbation Theory," *J. of Nuclear Science and Eng.* **62**, 9–19 (1977).

3.20. ASATANI, K., "Near Optimum Control of Distributed Parameter Systems Via Singular Perturbation Theory," *J. Math. Analysis and App.* **54**, 799–809 (1976).

3.21. AVRAMOVIC, B., "Iterative Algorithms for the Time Scale Operation of Large Scale Systems," Proceedings of the System Engineering for Power Organization: Forms for Large Scale Systems Conference, Davos, Switzerland (1979).

3.22. COTTON, F. A., Chemical Applications of Group Theory, Wiley-Interscience, New York, 1971.

3.23. HAMERMESH, M., *Group Theory*, Addison-Wesley, Reading, Massachusetts, 1962.

3.24. JAFFE, H. H., and MILTON, O., *Symmetry in Chemistry*, Wiley-Interscience, New York, 1965.

3.25. KOKOTOVIC, P. V., and YACKEL, R. A., "Singular Perturbation of Linear Regulators: Basic Theorems," *IEEE Transactions on Automatic Control*, **AC-17**, 29–37 (1972).

3.26. KOKOTOVIC, P. V., "An Introduction to Singular Perturbations," Proceedings of the Systems Engineering for Power: Organizational forms for Large Scale Systems Conference, Davos, Switzerland (1979).

3.27. LAMONT, J. S., *Applications of Finite Groups*, Academic Press, New York, 1959.

3.28. NIEVA, R., and CHRISTENSEN, G. S., "Symmetry Reduction of Reactor Systems," *Nuclear Science and Engineering* **64**, 791–795 (1977).

3.29. NIEVA, R., and CHRISTENSEN, G. S., "Symmetry Reduction of Linear Distributed Parameter Systems," *Int. J. Control* **36**(1), 143–153 (1982).

3.30. BOBONE, R., "A Computerized Solution of the Diffusion Equations by the Method of Solution Functions," *Transactions of the American Nuclear Society* **9**, 473 (1966).

3.31. BOBONE, R., "The Method of Solution Functions Extended from $r - \sigma$ to $r - z$ and $r - \sigma - z$ Geometry," *Transactions of the American Nuclear Society* **10**, 548 (1967).

3.32. BOBONE, R., "The Method of Solution Functions: A Computer-Oriented Solution of Boundary Value Problems as Applied to Nuclear Reactors," *Nuclear Science Engineering* **29**, 337 (1967).

4

Optimal Control of Distributed Nuclear Reactors

4.1. Introduction

This chapter is devoted to the problem of controlling the neutron flux distribution in a nuclear reactor core in which the spatial kinetic effects are important (Refs. 4.1 and 4.3). The problem consists of computing the control function that transfers the state of the system from an initial condition to a desired state, in a given period of time, and minimizes a quadratic performance index that penalizes the deviations from equilibrium so as to avoid potential spatial instabilities and high-power density spots (Refs. 4.17–4.20).

In Section 4.3, the optimization technique of the minimum norm problem in Hilbert spaces is applied to the problem of adjusting the neutron flux for a general distributed nuclear reactor whose dynamic behavior is described in the neighborhood of an equilibrium condition. In Section 4.4, the maximum principle is applied to a linear deterministic mathematical model of a nuclear reactor system. We also study the deviation of the approximation of this distributed system by discretization into a lumped system. Examples concerning the regulator and servomechanism problems of a regular reactor system are presented.

Kliger, in 1965 (Ref. 4.4), considered the problem of changing the neutron density level in such a way as to keep its spatial shape undistorted. The performance index to be minimized was the mean square error in space and time, between the actual system response and the desired trajectory. The reactor core was modeled by a one-energy neutron-group model with delayed neutron precursors. The model's parameters were assumed to be independent of the spatial coordinate, and a pseudocontrol function proportional to the product between the neutron multiplication factor and the neutron density was treated as a distributed control. The system state and the control were expanded in terms of an orthonormal set of spatial modes.

With the spatial independence of the system parameters, a decoupled set of ordinary differential equations was obtained for the expansion coefficients. The control term for each one of the spatial modes was determined by making use of known results for point reactor models. The distributed pseudocontrol effect thus determined was later approximated by a finite number of spatially concentrated control rods. In spite of its simplicity, Kliger's model (Ref. 4.4) introduced two features that have been used by many authors and, in a sense, have become standard techniques: it used an orthonormal expansion; and it treated a pseudocontrol function as a distributed input (Refs. 4.5–4.9).

In 1966 Wiberg (Refs. 4.10–4.12) studied the optimal feedback control of spatial xenon oscillations. He considered a general, linear reactor model and used a modal expansion in terms of the Kaplan modes. The finality property that characterizes this set of modes allowed for the decoupling of the power control system from the spatial regulator. The spatially concentrated control functions were modeled by means of modes and a quadratic performance index, which penalized the state deviations from equilibrium and measured the control energy. The control problem was reduced to manageable terms for optimal control by state-space methods. Wiberg's major contribution (Ref. 4.10) is the idea of decoupling the power control system from the spatial regulator system. His chapter has become one of the most important references in the field.

In 1967 Hsu and Bailey (Ref. 4.2) applied an extended version of the maximum principle to a simple one-group model for a homogeneous slab reactor with one delayed neutron precursor model. Their approach, based on Wang's work (Ref. 4.9) on distributed parameter systems, considered a quadratic cost functional.

Stacey (Ref. 4.7) applied several optimization techniques to the problem of controlling xenon spatial transients in thermal reactor cores. In 1968 (Ref. 4.7), he treated the problem in terms of dynamic programming formalism and gave a numerical example for a fairly realistic three-dimensional reactor model. The large number of state variables present, however, made it necessary to severely limit the set of allowable controls.

In 1969, Stacey (Ref. 4.7) applied the calculus of variations to distributed parameter systems to the xenon oscillation problem. In this case the optimality conditions were obtained in the form of a system of partial differential equations with mixed boundary conditions, which are very difficult to solve. A numerical example was given for the one-group model with xenon and iodine dynamics. A distributed control input and a quadratic cost functional were considered in the example, and an approximate solution was obtained by applying the methods of quasilinearization and nodal approximation. Knowing the large difference in the time constants of the neutron kinetics and the reactor poisoning process, Stacey (Ref. 4.7) assumed that changes in the neutron flux and temperature distributions occur instantaneously.

In 1970, he (Ref. 4.8) considered the application of variational synthesis to the optimal control of spatially independent reactor models. The method of variational synthesis consists of expanding the system state and the control input in terms of known functions of space and time. The conditions that minimize a given performance index are obtained by simple calculus in the form of algebraic equations. The choice of expansion functions in this case is arbitrary, and the approach lacks the rigor of other methods (Refs. 4.11–4.13).

Chaudhuri (Ref. 4.1), in 1972, applied the maximum principle to the problem of controlling xenon transients. He considered a quadratic cost functional and assumed a distributed control input. Solutions were obtained for a one-dimensional reactor model via space and space-time discretizations.

Lazarevich *et al.* (Ref. 4.5), in 1972, considered the application of dynamic programming methods to the problem of controlling the flux distribution. They obtained results for a slab reactor by applying Galerkin's method to the Hamilton-Jacobi canonical equations. An algorithm for computing the optimal control by using the eigenvectors of the Hamilton-Jacobi operator was also proposed. The approach is equivalent to Kaplan's modal expansion method.

The first application of the abstract formalism of functional analysis to the problem of controlling the neutron flux distribution in nuclear reactor cores was published by Kyong in 1968 (Ref. 4.21). Kyong treated the terminal state control problem for a reactor core of cylindrical configuration containing a finite number of control rods, with the neutron kinetics modeled by the one-group neutron diffusion equation. The unique optimal control for the terminal problem was shown to satisfy a coupled set of Fredholm integral equations of the second kind with degenerate kernels. A rigorous method, based on the characteristic expansions, was proposed for solving the integral equations.

It became clear from Kyong's work (Ref. 5.21) that the functional-analytic formulation yields necessary and sufficient conditions for optimality in a form amenable to the application of a different class of computational techniques, which in some cases may prove to be superior to the more familiar methods associated with the variational formulation of optimal control problems. These techniques approach the control problem through modal expansion methods and invoke variational principles that yield necessary conditions for optimality in the form of an infinite system of ordinary differential equations with mixed boundary conditions. In contrast, the functional-analytic approach yields necessary and sufficient conditions for optimality in the form of a finite set of integral equations.

4.2. The Reactor Core Model

The neutron dynamics for a reactor near its equilibrium state are described by a linear vector partial differential equation of the form

$$\frac{\partial}{\partial t}\psi(r, t) = A(r)\psi(r, t) + B(r)U(t) \tag{4.1}$$

where r is, in general, a three-dimensional spatial variable, defined on the reactor extrapolated volume. $\psi(r, t)$, for fixed r and t, is an N-dimensional vector containing the neutron flux at different energies, delayed neutron precursor densities, fission product and precursor densities, and fuel and moderator temperatures; $A(r)$ is a matrix spatial operator involving the gradient and Laplacian operators. For fixed time t, $U(t)$ is an M-dimensional vector representing the effect induced by the control rods; $B(r)$ is a rectangular matrix spatial operator of the appropriate dimensions.

Associated with equation (4.1) are the boundary conditions

$$\psi(r, t) = 0 \tag{4.2}$$

at the reactor extrapolated boundary and the initial condition

$$\psi(r, t_0) = Z_0(r)$$

at time t_0. The neutron flux and current must also satisfy continuity conditions at internal boundaries.

In general, we want to find a solution to equation (4.1) in a normed function space of interest. We consider the case in which, for each fixed time t, the state $\psi(r, t)$ is a function in a real Hilbert space H endowed with an inner product of the form

$$\langle Z(t), W(t) \rangle_H = \int_V W^T(r, t)Q(r)Z(r, t)\, dr \tag{4.3}$$

in which the integral is over the core extrapolated volume V. The term $Q(r)$ is an N-dimensional, positive-definite matrix with space-dependent entries, and W^T denotes the transpose of W.

Similarly, the control function $U(t)$, for fixed t, is assumed to be an element of a finite-dimensional, real Hilbert space E with inner product

$$\langle U(t), Y(t) \rangle_E = Y^T(t)RU(t) \tag{4.4}$$

where Y^T is the transpose of Y, and R is a positive-definite matrix of appropriate dimension. The control functions considered belong to the class of function with a bounded norm of the form

$$\int_{t_0}^{t_1} U^T(t)RU(t)\, dt = \int_{t_0}^{t_1} \|U(t)\|_E\, dt$$

We assume that equation (4.1) and the associated boundary conditions constitute a well-posed abstract Cauchy problem in relation to the norm topologies of H and E. Therefore, the unique solution to (4.1) and (4.2) can be written in the form

$$\psi(r, t) = G(r, t; r', t_0)Z_0(r') + \int_{t_0}^{t} G(r, t; r', \tau)B(r')U(\sigma)\, d\tau \quad (4.5)$$

where $G(r, t; r', \tau)$ is the strongly continuous semigroup associated with the operator $A(r)$, which satisfies

$$G(r, t_1; r', \tau)G(r, t_2; r', \tau) = G(r, t_1 + t_2; r', \tau)$$

$$G(r, t; r', t) = \text{identity}$$

and is also a strong solution to

$$\frac{\partial}{\partial t} G(r, t; r', \tau)\psi(r', \tau) = G(r, t; r', \tau)A(r')\psi(r', \tau)$$

in the Hilbert space H.

Equation (4.5) is the solution to (4.1) in the sense that ψ satisfies the relations

$$\lim_{h \to 0_+} \|h^{-1}[\psi(t + h) - \psi(t)] - A\psi(t) - BU(t)\|_H = 0$$

for all $t \geq t_0$, where $h \to 0_+$ denotes $h \to 0$ through positive values, and

$$\lim_{t \to 0_+} \|\psi(t) - Z\|_H = 0$$

for all initial conditions Z_0 in the domain of A;

$$D[A(r)] = \{Z(r); \|A(r)Z(r)\|_H < \infty, \text{ and } Z(r) = 0 \text{ in the boundary of } V\}$$

When the matrices Q and R, appearing in the inner-product definitions (4.3) and (4.4), are chosen to be the identity matrices in their corresponding spaces, it is clear that the solution (4.5) would satisfy (4.1) in the mean square sense and that the control functions being considered would belong to the class of controls with limited energy. The inclusion of a matrix Q other than the identity matrix provides a wider range of norms, which allows us, for example, to emphasize the difference between the state distributions and the equilibrium condition in a given region of the reactor, where core damage, due to large flux deviations, is more likely to occur.

Finally, a few explanatory remarks on the notation are mandatory. The semigroup $G(r, t; t', \tau)$ in our case represents an integral operator. The kernel of this operator depends on the variables r, t, t', τ, of which only r' defines the dummy variable of integration.

4.3. The Optimal Control Problem

The problem is to find the control vector $U(t)$ that brings the state distribution from an initial state to a desired equilibrium condition while minimizing the following performance index, which penalizes the deviations from equilibrium along the trajectory and the required control effort:

$$J(U) = \int_{t_0}^{t_1} \int_V [\psi(r, t) - Z(r)]^T Q(r) [\psi(r, t) - Z(r)]\, dr\, dt$$

$$+ k_0 \int_{t_0}^{t_1} U^T(t) R U(t)\, dt \qquad (4.6)$$

where Z is the desired state, k_0 is a positive constant chosen so as to establish the relative weight of the second term in (4.6), ψ is the system's state given by equation (4.5), and Q and R are defined as before.

It is convenient to introduce the following notation: H_1 and H_2 will denote Hilbert spaces with inner products

$$\langle U, Y \rangle_{H_1} = \int_{t_0}^{t_1} \langle U(t), Y(t) \rangle_E\, dt \qquad (4.7)$$

and

$$\langle Z, W \rangle_{H_2} = \int_{t_0}^{t_1} \langle Z(t), W(t) \rangle_H\, dt \qquad (4.8)$$

respectively. The operator F mapping H_1 into H_2 is defined by

$$F(r, t; \tau) U(\tau) = \int_{t_0}^{t} G(r, t; r', \tau) B(r') U(\tau)\, d\tau \qquad (4.9)$$

By using definitions (4.7) to (4.9), we can write the cost functional in equation (4.6) as

$$J(U) = \| FU - Z + G_0 Z_0 \|_{H_2} + k_0 \| U \|_{H_1} \qquad (4.10)$$

where Z_0 is the state distribution at time t_0 and G_0 is the semigroup defined by

$$G_0 = G(r, t; r', t_0)$$

which maps the space H_2 into itself.

4.3.1. A Minimum Norm Approach

The cost functional formulated in equation (4.10) is the same problem as problem 2 of Chapter 2. Therefore, the optimal control vector $U(t)$ can be determined from

$$[F^*F + K_0 I]U = F^*[Z - G_0 Z_0] \tag{4.11}$$

where F^* is the operator adjoint of F, which can be determined as follows.

4.3.1.1. The Adjoint F^*

The adjoint F^* is related to F through the inner-product relation

$$\langle FU, Z \rangle_{H_2} = \langle U, F^*Z \rangle_{H_1} \tag{4.12}$$

The left side of (4.12) can be expanded as

$$\langle FU, Z \rangle_{H_2} = \int_{t_0}^{t_1} \left\langle \int_{t_0}^{t} G(t; \tau) BU(\tau), Z(t) \right\rangle_H dt \tag{4.13}$$

where, for simplicity, only the temporal dependency is explicitly shown. The inner product in H involves an integral operation throughout the reactor core volume. After interchanging the order of integration, we can write (4.13) in the form

$$\langle FU, Z \rangle_{H_2} = \int_{t_0}^{t_1} \int_{t_0}^{t} \langle G(t; \tau) BU(\tau), Z(t) \rangle_H d\tau \, dt$$

which, in terms of the adjoint operators G^* and B^*, becomes

$$\langle FU, Z \rangle_H = \int_{t_0}^{t_1} \int_{t_0}^{t} \langle U(\tau), B^*G^*(t; \tau)Z(t) \rangle_E d\tau \, dt$$

After interchanging the order of integration, we have

$$\langle FU, Z \rangle_{H_2} = \int_{t_0}^{t_1} \int_{\tau}^{t_1} \langle U(\tau), B^*G^*(t; \tau)Z(t) \rangle_E dt \, d\tau$$

$$= \int_{t_0}^{t_1} \left\langle U(\tau), \int_{\tau}^{t_1} B^*G^*(t; \tau)Z(t) \, dt \right\rangle_E d\tau$$

from which

$$F^*(\tau, t)Z(t) = \int_{\tau}^{t_1} B^*G^*(t; \tau)Z(t) \, dt \tag{4.14}$$

Similarly, B^* is related to B by the relation

$$\langle BU, Z \rangle_H = \langle U, B^*Z \rangle_E \tag{4.15}$$

Expanding the left side of (4.15), we get

$$\langle BU, Z \rangle_H = \int_V Z^T(r)Q(r)B(r)U\,dv$$

$$= \int_V [B^T(r)Q^T(r)Z(r)]^T\,dv\,U$$

$$= \left[R^{-1} \int_V B^T(r)Q^T(r)Z(r)\,dv \right]^T RU$$

$$= \left\langle U, R^{-1} \int_V B^T(r)Q^T(r)Z(r)\,dv \right\rangle_E$$

from which it follows that

$$B^* = R^{-1} \int_V B^T(r)Q^T(r)Z(r)\,dv$$

where R has been assumed symmetric.

4.4. Necessary and Sufficient Conditions for Optimality: A Fredholm Integral Equation

We show that F^*F is an integral operator with nondegenerate kernel and that condition (4.12) constitutes a Fredholm equation. From equations (4.9) and (4.14), we find that the term F^*FU can be written as

$$F^*FU = \int_\tau^{t_1} B^*G(t; \tau) \int_{t_0}^t G(t; \alpha)BU(\alpha)\,d\alpha\,dt \qquad (4.16)$$

where, again, for simplicity only the temporal dependency is explicitly shown.

Interchanging the order of integration in expression (4.16), we get

$$F^*FU = \int_{t_0}^\tau \int_\tau^{t_1} B^*G^*(t; \tau)G(t; \alpha)B\,dt\,U(\alpha)\,d\alpha$$

$$+ \int_\tau^{t_1} \int_\alpha^{t_1} B^*G^*(t; \tau)G(t; \alpha)B\,dt\,U(\alpha)\,d\alpha$$

which, in turn, can be rewritten in the form

$$F^*FU = \int_{t_0}^{t_1} K(\tau; \alpha)U(\alpha)\,d\alpha$$

where

$$
K(\tau, \alpha) = \begin{cases} \displaystyle\int_{\tau}^{t_1} B^* G^*(t, \tau) G(t; \alpha) B \, dt & \text{for } \alpha < \tau \\[2ex] \displaystyle\int_{\alpha}^{t_1} B^* G^*(t; \tau) G(t; \alpha) B \, dt & \text{for } \alpha \geq \tau \end{cases} \tag{4.17}
$$

It follows from expressions (4.12) and (4.13) that the optimality condition is given by a Fredholm integral equation of the second kind:

$$
U(\tau) = \Delta(\tau) - \frac{1}{k_0} \int_{t_0}^{t_1} K(\tau; \alpha) U(\alpha) \, d\alpha \tag{4.18}
$$

where the function Δ is defined by

$$
\Delta(\tau) = \frac{1}{k_0} F^*[Z - G_0 Z_0]
$$

$$
= \frac{1}{k_0} \int_{\tau}^{t_1} B^* G^*(t; \tau)[Z - G_0 Z_0] \, dt \tag{4.19}
$$

From relation (4.17) note that $K(\tau, \alpha)$ is not in the form of a product or a sum of products of functions that depend on a single variable. In other words, $K(\tau, \alpha)$ is a nondegenerate kernel.

The necessary and sufficient conditions for optimality derived by Kyong for the terminal control problem are in the form of an integral equation with a degenerate kernel. Since the key ingredient in the characteristic expansion method proposed by Kyong (Ref. 4.21) for solving the optimality conditions is precisely the degeneracy of the kernel, it follows that this approach is not applicable to equation (4.18). However, equation (4.18) is amenable to the application of approximating techniques, such as contraction mapping algorithms or methods that reduce the integral equation into a system of algebraic equations via temporal discretization or function-expansion techniques.

4.5. Example: One-Neutron-Group Diffusion Equation

In this section we explain the application of the technique, described in the previous sections, through an example. Consider a homogeneous slab reactor model and the one-neutron-group diffusion equation

$$
\frac{\partial \psi(r, t)}{\partial t} = VD \left[\frac{\partial^2 \psi(r, t)}{\partial r^2} \right] + V(v\Sigma_f - \Sigma_a)\psi(r, t)
$$

$$
- V \sum_{i=1}^{M} u_i(t)\phi_0(r)\delta(r - r_i)
$$

with initial boundary conditions

$$\psi(r, t_0) = Z_r(r)$$

$$\psi(0, t) = \psi(b, t) = 0$$

Here, as usual, D is the neutron diffusion coefficient, Σ_f and Σ_a denote the fission and absorption macroscopic cross sections, respectively, V is the average neutron velocity, v denotes the number of neutrons generated per nuclear fission, and b denotes the reactor width. The control absorption cross section at location r_i is represented by $u_i(t)$.

The cost functional to be minimized by the control functions u_i is given by

$$J(U) = \int_{t_0}^{t_1} \int_0^b [\phi(r, t) - \phi_0(r)]^2 \, dr \, dt + k_0 \int_{t_0}^{t_1} \sum_{i=1}^M u_i^2(t) \, dt \quad (4.20)$$

Equation (4.20) penalizes the flux deviations from the equilibrium distribution $\phi_0(r)$, where

$$\phi_0(r) = \phi_M \sqrt{\frac{2}{b}} \sin\left(\frac{\pi r}{b}\right) \quad (4.21)$$

where ϕ_M is a constant. Equation (4.21) is a solution to

$$0 = D \frac{\partial^2}{\partial r^2} \phi(r) + [v\Sigma_f - \Sigma_a]\phi(r)$$

subject to the boundary condition.

Small flux deviations from $\phi_0(r)$,

$$\psi(r, t) = \phi(r, t) - \phi_0(r)$$

satisfy the linearized equation

$$\frac{\partial \psi}{\partial t}(r, t) = VD \frac{\partial^2}{\partial r^2} \psi(r, t) + V[v\Sigma_f - \Sigma_a]\psi(r, t)$$

$$- V \sum_{i=1}^M u_i(t)\phi_0(r)\delta(r - r_i) \quad (4.22)$$

with initial condition

$$\psi(r, t_0) = Z_0(r) = \phi(r, t) - \phi_0(r)$$

and boundary conditions

$$\psi(0, t) = \psi(b, t) = 0$$

Equation (4.22) can be solved by applying the method of separation of variables. The solution is

$$\psi(r, t) = \sum_n \frac{2}{b} \sin\left(\frac{n\pi r}{b}\right) \left[\int_0^b \sin\left(\frac{n\pi r'}{b}\right) Z_0(r') \, dr'\right] \exp \lambda_n(t - t_0)$$

$$+ \left[\frac{2}{b}\right]^{3/2} V\phi_M \int_{t_0}^t \sum_n \exp\{\lambda_n(t - \tau)\} \sin\left(\frac{n\pi r}{b}\right)$$

$$\times \left[\sum_{i=1}^M \sin\left(\frac{n\pi r_i}{b}\right) \sin\left(\frac{\pi r_i}{b}\right) u_i(\tau)\right] d\tau \qquad (4.23)$$

where

$$\lambda_n = \left[v\Sigma_f - \Sigma_a - \left(\frac{n\pi}{b}\right)^2 D\right] V \qquad (4.24)$$

In terms of the notation introduced at the beginning of this section, the solution (4.23) becomes

$$\psi(r, t) = G(r, t; r', t_0)Z_0(r') + \int_{t_0}^t G(r, t; r', \tau)B(r')U(\tau) \, d\tau$$

where

$$G(r, t; r', t_0)Z_0(r') = \sum_n \frac{2}{b} \sin\left(\frac{n\pi r}{b}\right) \left[\int_0^b \sin\left(\frac{n\pi r'}{b}\right) Z_0(r') \, dr'\right] \exp \lambda_n(t - t_0)$$

$$(4.25)$$

and

$$F(r, t; \tau)U(\tau)$$

$$= \left[\frac{2}{b}\right]^{3/2} V\phi_M \int_{t_0}^t \sum_n \exp\{\lambda_n(t - \tau)\} \sin\left(\frac{n\pi r}{b}\right) \left[\sum_{i=1}^M \sin\left(\frac{n\pi r_i}{b}\right) u_i(\tau)\right] d\tau$$

$$(4.26)$$

The operator $G(r, t; r', t_0)$, in this case, is self-adjoint. The adjoint F^* is given by an M-dimensional vector operator with entries $f_i^*(\tau; r, t)$ defined by

$$f_i^*(\tau; r, t)Z(r, t)$$

$$= V\phi_M \left(\frac{2}{b}\right)^{3/2} \int_\tau^{t_1} \int_0^b \sum_n \exp\{\lambda_n(t - \tau)\} \left[\sin\left(\frac{n\pi r_i}{b}\right) \sin\left(\frac{\pi r_i}{b}\right)\right]$$

$$\times \sin\left(\frac{n\pi r}{b}\right) Z(r, t) \, dr \, dt$$

$$i = 1, 2, \ldots, M \qquad (4.27)$$

Using (4.26) and (4.27), we find that the ith entry in F^*FU takes the form

$$f_i^*(\tau; r, t) F(r, t; \alpha) U(\tau)$$

$$= \frac{2}{b} \int_\tau^{t_1} \int_0^b \sum_n \exp\{\lambda_n(t - \tau)\} b_{ni} \sin\left(\frac{n\pi r}{b}\right)$$

$$\times \int_{t_0}^t \sum_m \exp\{\lambda_m(t - \tau)\} \sin\left(\frac{m\pi r}{b}\right) \sum_{j=1}^M b_{mn} u_j(\alpha) \, d\alpha \, dr \, dt$$

where

$$b_{ni} = \sin\left(\frac{n\pi r_i}{b}\right) \sin\left(\frac{\pi r_i}{b}\right) \phi_M\left(\frac{2}{b}\right) V \tag{4.28}$$

and, given the orthogonality property,

$$\frac{2}{b} \int_0^b \sin\left(\frac{n\pi r}{b}\right) \sin\left(\frac{m\pi r}{b}\right) dr = \begin{cases} 1 & \text{for } n = m \\ 0 & \text{for } n \neq m \end{cases}$$

expression (4.11) becomes

$$f^*(\tau; r, t) F(r, t; \alpha) U(\alpha)$$

$$= \int_\tau^{t_1} \int_{t_0}^t \sum_n \exp\{\lambda_n(2t - \tau - \alpha)\} b_{ni} \sum_{j=1}^M b_{nj} u_j(\alpha) \, d\alpha \, dt$$

which after interchanging the order of integration takes the form

$$f_i^*(\tau; r, t) F(r, t; \alpha) U(\alpha) = \int_{t_0}^{t_1} \sum_{j=1}^M \sum_n b_{ni} b_{nj} K_n(\tau; \alpha) u_j(\alpha) \, d\alpha \tag{4.29}$$

where

$$K_n(\tau; \alpha) = \begin{cases} \int_\tau^{t_1} \exp \lambda_n(2t - \tau - \alpha) \, dt & \text{for } \alpha < \tau \\ \int_\alpha^{t_1} \exp \lambda_n(2t - \tau - \alpha) \, dt & \text{for } \alpha \geq \tau \end{cases}$$

and upon integration

$$K_n(\tau; \alpha) = \begin{cases} [\exp \lambda_n(2t_1 - \tau - \alpha) \\ \quad - \exp \lambda_n|\tau - \alpha|]/2\lambda_n & \text{for } \lambda_n \neq 0 \\ t_1 - \tau & \text{for } \lambda_n = 0 \text{ and } \alpha < \tau \\ t_1 - \alpha & \text{for } \lambda_n = 0 \text{ and } \alpha \geq \tau \end{cases} \tag{4.30}$$

where $|\cdot|$ denotes absolute value.

The function $\Delta(\tau)$ defined in (4.19) becomes

$$\Delta(\tau) = -\frac{1}{k_0} F^*(\tau; r, t) G(r, t; r', t_0) Z_0(r')$$

and, using (4.29) and (4.25), we obtain the ith entry $\Delta_i(\tau)$:

$$\Delta_i(\tau) = -\frac{1}{k_0}\left[\frac{2}{b}\right]\int_\tau^{t_1}\int_0^b \sum_n \exp\{\lambda_n(t-\tau)\}b_{ni}\sin\left(\frac{n\pi r}{b}\right)\sum_m \frac{2}{b}\sin\left(\frac{m\pi r}{b}\right)$$

$$\times \left[\int_0^b \sin\left(\frac{m\pi r'}{b}\right)Z_0(r')\,dr'\right]\exp\lambda_m(t-t_0)\,dr\,dt$$

which reduces to

$$\Delta_i(\tau) = -\frac{1}{k_0}\left[\frac{2}{b}\right]\sum_n\left[\int_\tau^{t_1}\exp\lambda_n(2t-\tau-t_0)\,dt\right]b_{ni}\int_0^b\sin\left(\frac{n\pi r'}{b}\right)Z_0(r')\,dr'$$

and becomes, after integration,

$$\Delta_i(\tau) = \frac{1}{k_0 b}\sum_n\frac{\exp\lambda_n(\tau-t_0)-\exp\lambda_n(2t_1-t_0-\tau)}{\lambda_n}$$

$$\times b_{ni}\int_0^b\sin\left(\frac{n\pi r'}{b}\right)Z_0(r')\,dr' \tag{4.31}$$

The optimality conditions are finally obtained in the form

$$u_i(\tau) = \Delta_i(\tau) - \frac{1}{k_0}\int_{t_0}^{t_1}\sum_{j=1}^M \hat{K}_{ij}(\tau;\alpha)u_j(\alpha)\,d\alpha, \qquad i=1,2,\dots,M \tag{4.32}$$

where $\hat{K}_{ij}$ is defined by

$$\hat{K}_{ij} = \sum_n b_{ni}b_{nj}K_n(\tau;\alpha) \tag{4.33}$$

4.6. Discussion

More conventional techniques approach this control problem through modal expansion methods and invoke variational principles that yield necessary conditions for optimality in the form of an infinite system of ordinary differential equations with mixed boundary conditions. The question of how many modes should be included in a finite-dimensional version of the system is a very difficult problem, which in most cases can be solved only by trial and error. In contrast, condition (4.32) forms a system of M integral equations with nondegenerate kernels. Although these kernels are represented by infinite series, the question of how many terms should be considered in a finite series is a tractable problem and, as shown below, error bounds can be estimated with relative ease.

Consider the kernel

$$\hat{K}_{ij}(\tau;\alpha) = \sum_{n=1}^{\infty} b_{ni}b_{nj}K_n(\tau;\alpha)$$

and the approximate version

$$\hat{K}_{ij}(\tau; \alpha) = \sum_{n=1}^{P} b_{ni} b_{nj} K_n(\tau; \alpha)$$

which is obtained by truncating the series after the first P terms. The error is given by

$$\varepsilon_{ij}(\tau; \alpha) = \hat{K}_{ij}(\tau; \alpha) - \tilde{K}_{ij}(\tau; \alpha)$$

$$= \sum_{n=P+1}^{\infty} b_{ni} b_{nj} K_n(\tau; \alpha)$$

where $K_n(\tau; \alpha)$ is defined by

$$K_n(\tau; \alpha) = \frac{\exp \lambda_n(2t_1 - \tau - \alpha) - \exp \lambda_n |\tau - \alpha|}{2\lambda_n}$$

In view of equation (4.28), the product $b_{ni} b_{nj}$ is bounded absolutely by

$$|b_{ni} b_{nj}| \le \left[\frac{2}{b} \phi_M V \right]^2$$

Therefore

$$|\varepsilon_{ij}(\tau; \alpha)| \le \left[\frac{2}{b} \right]^2 \phi_M^2 V^2 \sum_{n=P+1}^{\infty} |K_n(\tau; \alpha)|$$

and, by (4.30), it follows that

$$|\varepsilon_{ij}(\tau; \alpha)| < \left[\frac{\phi_M V}{b} \right]^2 2 \sum_{n=P+1}^{\infty} \frac{\exp \lambda_n(2t_1 - \tau - \alpha) + \exp \lambda_n |\tau - \alpha|}{|\lambda_n|} \quad (4.34)$$

for all subscripts i and j. After integrating the strict inequality (4.34), we have

$$\int_{t_0}^{t_1} \int_{t_0}^{t_1} |\varepsilon_{ij}(\tau; \alpha)| \, d\alpha \, d\tau$$

$$< \left[\frac{\phi_M V}{b} \right]^2 2 \sum_{n=P+1}^{\infty} \left[\frac{2[t_1 - t_0]}{[\lambda_n]^2} + \frac{\exp \lambda_n 2(t_1 - t_0) - 1}{|\lambda_n|^3} \right] \quad (4.35)$$

Recall from (4.24) that

$$\lambda_n = [a_1 - n^2] a_2 \quad (4.36)$$

where a_1 and a_2 are positive real constants given by

$$a_1 = \frac{v\Sigma_f - \Sigma_a}{\pi^2 D} b^2 \le 1$$

and

$$a_2 = \frac{V\pi^2 D}{b^2}$$

Using the inequality

$$\exp \lambda_n 2(t_1 - t_0) - 1 < 1$$

and substituting (4.36) in (4.35), we get

$$\int_{t_0}^{t_1} \int_{t_0}^{t_1} |\varepsilon_{ij}(\tau; \alpha)| \, d\sigma \, d\tau$$

$$< 2 \left[\frac{\phi_M V}{b} \right]^2 \sum_{n=p+1}^{\infty} \left[\frac{2[t_1 - t_0]}{[a_2]^2 [n^2 - 1]^2} + \frac{1}{[a_2]^3 [n^2 - 1]^3} \right]$$

Replacing summation by integration and evaluating the integral, we obtain the error bound

$$\int_{t_0}^{t_1} \int_{t_0}^{t_1} |\varepsilon_{ij}(\tau; \alpha)| \, d\alpha \, d\tau$$

$$< \frac{4\phi_M^2 b^2}{\pi^4 D^2} \frac{t_1 - t_0}{[P^2 + 2P]^2}$$

$$+ \frac{2\phi_M^2 b^2}{\pi^4 D^2} \left(2[t_1 - t_0] - \frac{3b^2}{4V\pi^2 D} \right) \left[\frac{P+1}{2P[P+2]} + \frac{1}{4} \ln \left(\frac{P}{P+2} \right) \right]$$

$$+ \frac{\phi_M^2 b^4}{2V\pi^6 D^3} \frac{[P+1][P^2 + 2P] + 3}{[P^2 + 2P]^3} \Bigg]$$

This estimate is a function of P, which represents the number of terms considered in the finite series approximation to $\hat{K}_{ij}(\tau; \alpha)$.

4.7. Method for Computing the Optimal Control

Rewrite equation (4.32) in a more compact form as

$$U(\tau) = L(\tau; \alpha) U(\alpha) \tag{4.37}$$

where $U(\tau)$ is an element of the space H_1 and L is a nonlinear transformation from H_1 into itself, defined by

$$L(\tau, \alpha) U(\alpha) = \Delta(\tau) - \frac{1}{k_0} \int_{t_0}^{t_1} K(\tau; \alpha) U(\alpha) \, d\alpha$$

and $\Delta(\tau)$ is a function in H_1 with entries $\Delta_i(\tau)$ defined by equation (4.31). $\hat{K}(\tau; \alpha)$ is a matrix operator with entries $\hat{K}_{ij}(\tau; \alpha)$ defined by equation (4.67).

Invoking the contraction mapping theorem, we see that the sequence

$$U_n(\tau) = L(\tau; \alpha) U_{n-1}(\alpha)$$

based on the initial guess $U_0(\tau)$ in H_1 would converge to the unique solution to equation (4.37) provided that L is a contraction; that is, for any functions U and W in H_1,

$$\|LU - LW\|_{H_1} \le C\|U - W\|_{H_1}$$

where C is a constant, $0 < C < 1$.

We will compute a rough estimate of C in order to test for the contraction condition and to establish a criterion for the applicability of the successive approximation algorithm to equation (4.37). Consider the norm

$$\left\| \int_{t_0}^{t_1} K(\tau; \alpha) U(\alpha) \, d\alpha \right\|_{H_1^2} \tag{4.38}$$

and for convenience denote (4.38) by d. Using the definition of $K(\tau; \alpha)$, we can expand d as

$$d = \int_{t_0}^{t_1} \int_{t_0}^{t_1} \int_{t_0}^{t_1} \sum_{i=1}^{M} \sum_{j=1}^{M} \sum_{k=1}^{M} \hat{K}_{ij}(\tau; \alpha)\hat{K}_{ik}(\tau; \beta)u_j(\alpha)u_k(\beta) \, d\alpha \, d\beta \, d\tau \tag{4.39}$$

where u_j is the jth component of U. Substituting the series representation (4.33) in (4.39) and taking the absolute value of the integrand, we get

$$d \le [\max_{ni} b_{ni}]^4 M \int_{t_0}^{t_1} \sum_{n=1}^{\infty} \left[\int_{t_0}^{t_1} |K_n(\tau; \alpha)| \sum_j |u_j(\alpha)| \alpha \right]^2 d\tau$$

which by Schwarz's inequality reduces to

$$d \le [\max_{ni} b_{ni}]^4 M \left[\sum_{n=1}^{\infty} \int_{t_0}^{t_1} \int_{t_0}^{t_1} [K_n(\tau, \alpha]^2 \, d\alpha \, d\tau \right] \|U\|_{H_1^2} \tag{4.40}$$

In view of relation (4.30), it is clear that

$$K_n^2(\tau; \alpha) \le \begin{cases} [\exp 2\lambda_n(2t_1 - \tau - \alpha) + \exp 2\lambda_n|\tau - \alpha|]/4[\lambda_n]^2, & \text{for } n > 1 \\ [t_1 - t_0]^2, & \text{for } n = 0 \end{cases}$$

Hence, after integration, we have

$$\int_{t_0}^{t_1} \int_{t_0}^{t_1} [K_n(\tau; \alpha)]^2 \, d\alpha \, d\tau$$

$$< \frac{\exp 4\lambda_n(t_1 - t_0) - \exp 2\lambda_n(t_1 - t_0) + \exp \lambda_n(t_1 - t_0) - 1}{16[\lambda_n]^4} - \frac{[t_1 - t_0]}{4[\lambda_n]^3}$$

$$\text{for } n > 1$$

Since λ_n is negative for all $n > 1$, it follows that

$$\int_{t_0}^{t_1} \int_{t_0}^{t_1} [K_n(\tau; \alpha)]^2 \, d\alpha \, d\tau < \begin{cases} \dfrac{[t_1 - t_0]}{2|\lambda_n|^3}, & \text{for } n > 1 \\[2mm] [t_1 - t_0]^4, & \text{for } n = 1 \end{cases} \tag{4.41}$$

By the definition of λ_n (4.24), equation (4.41) reduces to

$$\sum_{n=1}^{\infty} \int_{t_0}^{t_1} \int_{t_0}^{t_1} [K_n(\tau; \alpha)]^2 \, d\alpha \, d\tau < [t_1 - t_0]^4 + \sum_{n=2}^{\infty} \frac{[t_1 - t_0]b^6}{(VD\pi^2)^3 2} \frac{1}{(n^2 - 1)^3}$$

From the equation

$$\sum_{n=2}^{\infty} \frac{1}{(n^2 - 1)^3} = 3.94 \times 10^{-2}$$

it follows that

$$\sum_{n=1}^{\infty} \int_{t_0}^{t_1} \int_{t_0}^{t_1} [K_n(\tau; \alpha)]^2 \, d\alpha \, d\tau < [t_1 - t_0]^4 + 5.77 \times 10^{-3} \frac{[t_1 - t_0]b^6}{(VD\pi^2)^3} \tag{4.42}$$

Using (4.42) and (4.40), we obtain the upper bound estimate

$$\left\| \int_{t_0}^{t_1} K(\tau; \alpha) U(\alpha) \, d\alpha \right\|_{H_1}$$

$$< \left[\frac{2\phi_M V}{b} \right]^2 M^{1/2} \left[[t_1 - t_0]^4 + 2.97 \times 10^{-1} \frac{[t_1 - t_0]b^6}{(VD\pi^2)^3} \right]^{1/2} \|U\|_{H_1}$$

Given that

$$\|LU - LW\|_{H_1} \leq \frac{1}{k_0} \left\| \int_{t_0}^{t_1} [K(\tau; \alpha) - W(\alpha)] \, d\alpha \right\|_{H_1}$$

it follows from (4.42) that

$$\|LU - LW\|_{H_1} < C \|U - W\|_{H_1}$$

where

$$C = \frac{1}{k_0} \left[\frac{2\phi_M V}{b} \right]^2 M^{1/2} \left[[t_1 - t_0]^4 + 1.97 \times 10^{-2} \frac{[t_1 - t_0]b^6}{(VD\pi^2)^3} \right]^{1/2} \tag{4.43}$$

It is clear from (4.43) that a sufficiently large value of k_0 would make C less than one. The design parameter k_0, introduced in the performance index (4.20), provides the flexibility required to obtain the desired balance between system response and control action. A suitable value of k_0 can only be found through systematic searching, which of necessity involves repeated computations for different values of k_0.

Assuming that the contribution of the flux deviation term at its maximum in the performance index should equal the contribution of the control energy term at its maximum, Wiberg outlined a method for obtaining an initial rough estimate of k_0. For the case where the flux deviation is to be no more than 10% of the steady-state flux and where the control reactivity is to be no more than 10% of the total material reactivity, Wiberg proposed the initial rough estimate

$$k_0 = \frac{2M}{b}\left[\frac{\phi_M}{v\Sigma_f - \Sigma_a}\right]^2 \tag{4.44}$$

where M is the number of controls, and the other variables are defined as before.

Substituting (4.44) in (4.43), we obtain

$$C = \frac{2}{b}[(v\Sigma_f - \Sigma_a)V]^2\left[\frac{[t_1 - t_0]^4}{M} + 1.97 \times 10^{-2}\frac{(t_1 - t_0)b^6}{M(VD\pi^2)^3}\right]^{1/2} \tag{4.45}$$

Although (4.45) is a very conservative estimate, it provides a useful criterion for determining if the contraction mapping algorithm would yield a convergent sequence. If $C < 1$ and U denotes the solution to equation (4.37), then the contraction mapping theorem ensures that the sequence

$$U_n = LU_{n-1}$$

converges, with rate of convergence

$$\|U - U_n\|_{H_1} \leq \frac{C^n}{1 - C}\|LU_0 - U_0\|_{H_1}$$

where U_0 is the initial guess. Consider, for example, the reactor constants given by Wiberg: $(v\Sigma_f - \Sigma_a)V = 0.256 \text{ sec}^{-1}$, $b = 250$ cm, $M = 2$, and $VD = 1600 \text{ cm}^2/\text{sec}$. In this case, the criterion (4.45) takes the form

$$C = 5.24 \times 10^{-4}[0.5[t_1 - t_0]^4 + 6.11 \times 10^{-1}[t_1 - t_0]]^{1/2}$$

and since the time scale of the nuclear model considered is of the order of seconds, it can be concluded that the successive approximation algorithm is applicable to problems in this range.

4.8. Maximum Principle Approach (Ref. 4.1)

In Section 4.7, the optimization technique of the minimum norm problem in Hilbert spaces was applied to the problem of adjusting the neutron flux for a general distributed nuclear reactor, whose dynamic behavior is described in the neighborhood of an equilibrium condition. In this section, the maximum principle is applied to a linear deterministic distributed nuclear reactor system. The optimal control laws for a distributed nuclear reactor system described by parabolic partial differential equations have been formulated through space discretization schemes. Also, studied are the stability and convergence of the discrete model described by difference equations.

For the one-group reactor model, the equation

$$\phi_t(Y, t) = D\nabla^2\phi(Y, t) - \Sigma_a V\phi(Y, t) + u(Y, t) \tag{4.46}$$

is used, with initial condition

$$\phi(Y, 0) = \phi_0(Y) \tag{4.47}$$

and boundary condition

$$\phi_Y(Y, t)_{\partial\Omega} = \alpha \tag{4.48}$$

where ϕ = thermal neutron flux, which is a function of time and the spatial coordinate vector Y

Y = n-dimensional vector $Y^T = [y_1, y_2, \ldots, y_n]$ belonging to the space Ω

$\partial\Omega$ = boundary of Ω

D = diffusion coefficient, given by $D = \frac{1}{3}V\lambda_{tr}$

λ_{tr} = transport mean free path of neutrons

Σ_a = macroscopic absorption cross section

V_2 = speed of neutrons

∇ = Laplacian operator in n-dimensional vector space

$\phi_0(Y)$ = arbitrary function of Y

α = n-dimensional constant vector

The xenon and iodine concentrations for the production of fission fragments of xenon-135 and iodine-135 are

$$I_t(Y, t) = -\lambda_I I(Y, t) + \gamma_I \Sigma_f \phi(Y, t) \tag{4.49}$$

$$X_t(Y, t) = \lambda_I I(Y, t) + \gamma_X \Sigma_f \phi(Y, t) - \lambda_X X(Y, t) - \sigma_X \phi(Y, T) \tag{4.50}$$

with initial condition

$$I(Y, 0) = I_0(Y) \tag{4.51}$$

$$X(Y, 0) = X_0(Y) \tag{4.52}$$

where $I(Y, t)$ = number of atoms of ^{135}I present per reactor volume of any space and time

λ_I, λ_X = decay constants of ^{135}I and ^{135}Xe, respectively

γ_I, γ_X = fractional yield of ^{135}I and ^{135}Xe, respectively

Σ_f = macroscopic fission cross section of fuel

σ_x = microscopic thermal neutron absorption cross section of ^{135}Xe

In this reactor model, the temperature–negative feedback effect of the reactor will be neglected. The nuclear reactor system model is given in Figure 4.1.

4.8.1. Problem Formulation

The problem of the nuclear reactor model given in Figure 4.1 is to find the control vector $u(Y, t)$ that minimizes the objective function

$$J = \int_\Omega \theta[\phi(Y, t)] \int_{t_0}^{t_f} d\Omega$$

$$+ \int_{t_0}^{t_f} \int_\Omega \{\Phi[Y, t, \phi(Y, t), \phi_Y(Y, t), \phi_{YY}(Y, t), u(Y, t)]\}\, d\Omega\, dt \qquad (4.53)$$

where the functions θ and Φ are scalar functions of the arguments shown, subject to satisfying the constraint given by equation (4.46).

Define the Hamiltonian H as

$$H[Y, t, \phi(Y, t)\phi_{YY}(Y, t), u(Y, t), \lambda(Y, t)]$$

$$= \Phi[Y, t\phi(Y, t), \phi_Y(Y, t), \sigma_{YY}(Y, t)u(Y, t)]$$

$$+ \lambda(Y, t)[D\nabla^2\phi(Y, t) - \Sigma_a V\phi(Y, t) + u(Y, t)]$$

By applying the maximum principle, we obtain the necessary conditions for optimization:

$$H_\phi + \lambda_t - \frac{\partial}{\partial Y}\left[\frac{\partial H}{\partial(\partial\phi/\partial Y)}\right] + \frac{\partial^2}{\partial Y^2}\left[\frac{\partial H}{\partial(\partial^2\phi/\partial Y^2)}\right] = 0 \qquad (4.54)$$

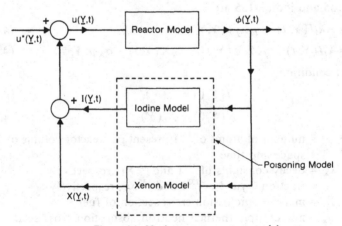

Figure 4.1. Nuclear reactor system model.

For no inequality constraint on the control vector,

$$H_u = 0 \tag{4.55}$$

The transversality condition is

$$\theta\big|_{t_0}^{t_f} = \lambda\big|_{t_0}^{t_f} \tag{4.56}$$

$$\frac{\partial}{\partial Y}\left[\frac{\partial H}{\partial(\partial^2\phi/\partial Y^2)}\right]_{\partial\Omega} = 0, \qquad \frac{\partial}{\partial Y}\left[\frac{\partial H}{\partial(\partial\phi/\partial Y)_{\partial\Omega}}\right] = 0 \tag{4.57}$$

To find the optimal control $u(Y, t)$, we solve equations (4.46), (4.54), and (4.55) with the two-point spatial boundary conditions of equations (4.47), (4.48), (4.56), and (4.57). After optimal values of $u(Y, t)$ and $\phi(Y, t)$ are known, we solve equations (4.49) and (4.50) with conditions (4.51) and (4.52) for $I(Y, t)$ and $X(Y, t)$ Hence, the input signal $u^*(Y, t)$ can be obtained. In the next section we discuss how to obtain $u(Y, t)$ for special problem formulations.

4.8.2. Regulator Problem

In this section, we offer formulations of the previous section for a one-dimensional linear distributed reactor system, where the objective function of equation (4.53) is of a regulator-type problem, minimizing a quadratic function of the state and the control vector.

The one-dimensional linear distributed reactor system can be modeled by

$$\phi_t(y, t) = D\phi_{yy}(y, t) - \Sigma_a V\phi(y, t) + u(y, t) \tag{4.58}$$

$$\phi(y, 0) = \phi_0(y) \tag{4.59}$$

$$\phi_y(0, t) = 0 \tag{4.60}$$

$$\phi_y(y_f, t) = 0$$

$$I_t(y, t) = -\lambda_I I(y, t) + \gamma_I \Sigma_f \phi(y, t) \tag{4.61}$$

$$X_t(y, t) = \lambda_I I(y, t) + \gamma_x \Sigma_f \phi(y, t) - \lambda_x x(y, t) - \sigma_x \phi(y, t) \tag{4.62}$$

$$I(y, 0) = I_0(y)$$

$$X(y, 0) = X_0(y)$$

where $\phi(y, t)$ is the thermal neutron flux at any time t and spatial coordinate axis y, $u(y, t)$ represents the unconstrained distributed control, and y_f is the final point on the spatial coordinate axis y.

The regulator problem can be stated as compute the control signal $u(y, t)$ that minimizes the quadratic cost function

$$J = \frac{1}{2} \int_0^{t_f} \int_0^{y_f} [Q\phi^2(y, t) + RU^2(y, t)] \, dy \, dt \qquad (4.63)$$

subject to satisfying the system constraint given by equation (4.58). In equation (4.63), Q and R are known constants and t_f and y_f are fixed.

Define the Hamiltonian H as

$$H[y, t, \phi(y, t), \phi_{yy}(y, t), u(y, t), \lambda(y, t)]$$
$$= Q\phi^2(y, t) + Ru^2(y, t) + \lambda(y, t)[D\phi_{yy}(y, t) - \Sigma_a V\phi(y, t) + U(y, t)]$$

By applying the maximum principle as before, we obtain the necessary conditions

$$Q\phi(y, t) - \lambda \Sigma_a V + \lambda_t(y, t) + D\lambda_{yy}(y, t) = 0 \qquad (4.64)$$

$$Ru(y, t) + \lambda(y, t) = 0 \qquad (4.65)$$

$$\lambda(y, t_f) = 0$$

$$\lambda_y(0, t) = 0$$

$$\lambda_y(y_f, t) = 0$$

If we substitute equations (4.64) and (4.65) into equation (4.58), we obtain a single partial differential equation in the control variable:

$$D^2 Ru_{yyyy}(y, t) - 2RD\Sigma_a Vu_{yy}(y, t) - Ru_{tt}(y, t) + [Q + (\Sigma_a V)^2 R]u(y, t) = 0$$
$$(4.66)$$

Solving this equation gives the control variable $u(y, t)$ if Q, R, D, Σ_a and V are known.

4.8.3. Servomechanism Problem

For this problem we use a one-dimensional linear distributed model given by equation (4.58), but the cost functional to be minimized is slightly different from that given by equation (4.63). The problem can be formulated as follows: derive the control signal $u(y, t)$ that minimizes the cost function

$$J = \frac{1}{2} \int_0^{t_f} \int_0^{y_f} [Q\{\phi(y, t) - \Psi(y, t)\}^2 + Ru^2(y, t)] \, dy \, dt \qquad (4.67)$$

and satisfies the equality constraints given by equation (4.58).

The Hamiltonian H can be defined as

$$H[y, t, \phi(y, t), \phi_{yy}(y, t), u(y, t), \lambda(y, t)]$$
$$= Q\{\phi(y, t) - \Psi(y, t)\}^2 + Ru^2(y, t)$$
$$+ \lambda(y, t)[D\phi_{yy}(y, t) - \Sigma_a V\phi(y, t) + u(y, t)]$$

By applying the maximum principle, we obtain necessary conditions

$$Q[\phi(y, t) - \Psi(y, t)] - \lambda \Sigma_a V + \lambda_t(y, t) + D\lambda_{yy}(y, t) = 0 \qquad (4.68)$$

$$Ru(y, t) + \lambda(y, t) = 0 \qquad (4.69)$$

$$\lambda(y, t_f) = 0$$

$$\lambda_y(0, t) = 0$$

$$\lambda_y(y_f, t) = 0$$

If we substitute equations (4.68) and (4.69) into equation (4.58), we obtain a single partial differential equation in the control variable:

$$D^2 Ru_{yyyy}(y, t) - 2RD\Sigma_a Vu_{yy}(y, t) - Ru_{tt}(y, t) + [Q + (\Sigma_a V)^2 R]u(y, t)$$

$$+ DQ\Psi_{yy}(y, t) - Q\Psi_t(y, t) - \Sigma_a VQ\Psi(y, t) = 0 \qquad (4.70)$$

Solving equation (4.70) gives the control variable $u(y, t)$ if Q, R, D, Σ_a, and V are known.

The solutions to equations (4.66) and (4.70) are very complex. Therefore, effective approximation schemes are required to transform the distributed parameter system into a lumped system. In the next sections we discuss two methods used to transform the distributed system to a discretized system — spatial discretization and space-time discretization. These methods are used to compute the optimal control law for the problems formulated above.

4.8.4. Spatial Discretization Scheme

If y_f is the final point on the spatial coordinate axis y, then the size of the space increment is

$$\Delta y = \frac{y_f}{n}$$

where n is an integer. From the central difference formula, the model given in equation (4.58) can be written in the spatially discrete version as

$$\frac{d\phi_i(t)}{dt} = [D/(\Delta y)^2] Y[\phi_{i-1}(t) - 2\phi_i(t) + \phi_{i+1}(t)] - \Sigma_a V\phi_i(t) + u_i(t)$$

$$i = 1, 2, \ldots, n \qquad (4.71)$$

The boundary conditions given by equations (4.59) and (4.60) can be written as

$$\phi_0 = \phi_1$$

and $\qquad (4.72)$

$$\phi_n = \phi_{n+1}$$

if we use forward and backward differences.

If we use equation (4.72) in equation (4.71), we can write equation (4.71) in compact vector form as

$$\frac{d\Phi(t)}{dt} = A\Phi(t) + BU(t) \tag{4.73}$$

and

$$\Phi(0) = \Phi_0$$

where

$$\Phi(t) = \text{col}(\phi_1(t), \ldots, \phi_n(t))$$
$$U(t) = \text{col}(u_1(t), \ldots, u_n(t))$$

and B is an nth-order unity matrix. Furthermore, A is an $n \times n$ tridiagonal matrix given by

$$A = \begin{bmatrix} -(p+r) & r & 0 & 0 & 0 & 0 \\ r & -(p+2r) & r & 0 & 0 & 0 \\ 0 & 0 & 0 & r & -(p+2r) & r \\ 0 & 0 & 0 & 0 & r & -(p+r) \end{bmatrix}$$

where we define

$$p = \Sigma_a V \tag{4.74}$$

and

$$r = \frac{D}{(\Delta y)^2} \tag{4.75}$$

Applying the same principle, we can write the dynamics of xenon and iodine described in equations (4.61) and (4.62) in a spatial discretization scheme as

$$\dot{I}(t) = -\lambda_I I(t) + \gamma_I \Sigma_f \Phi(t)$$
$$\dot{X}(t) = \lambda_I I(t) + \gamma_X \Sigma_f \phi(t) - \lambda_X X(t) - \sigma_X \Phi(t)$$

with boundary conditions

$$I(0) = I_0 \quad \text{and} \quad X(0) = X_0$$

for n-dimensional vector functions $I(t)$ and $X(t)$.

4.8.5. Space-Time Discretization Scheme

If m and n are integers, then the size of the space is

$$\Delta y = \frac{y_f}{m}$$

and the size of the incremental time is

$$\Delta t = \frac{t_f}{n}$$

In this case, the model of the system described in equation (4.58) can be written in the space-time discretization scheme as

$$\phi_{j,k+1} = \phi_{j,k} + \Delta t[\phi_{j+1,k}^{-2}\phi_{j,k} + \phi_{j-1,k}]r - p\Delta t\phi_{j,k} + \Delta t u_{j,k} \qquad (4.76)$$

where p and r are defined in equations (4.74) and (4.75). Using forward and backward differences, we can write the boundary conditions of equations (4.59) and (4.60) as

$$\phi_{0,k} = \phi_{-1,k} \qquad \text{and} \qquad \phi_{m,k} = \phi_{m+1,k}$$

Equation (4.76) can be written by a single vector difference equation as

$$\Phi(N+1) = A\Phi(N) + BU(N)$$

where $N = 0, 1, 2, \ldots, n$ and $\Phi(N)$ is an $(m+1)$-dimensional state vector given by

$$\Phi^{T}(N) = [\phi_{0,N}, \phi_{1,N}, \ldots, \phi_{m,N}]$$

$u(N)$ is an $(m+1)$-dimensional control vector given by

$$U^{T}(N) = [u_{0,N}, u_{1,N}, \ldots, u_{m,N}]$$

and

$$B = \Delta t I,$$

where I is an $(m+1) \times (m+1)$ unity matrix, and

$$A = \begin{bmatrix} a & b & 0 & 0 & 0 & 0 & 0 \\ b & c & b & 0 & 0 & 0 & 0 \\ 0 & b & c & b & 0 & 0 & 0 \\ 0 & 0 & 0 & 0 & b & c & b \\ 0 & 0 & 0 & 0 & 0 & b & c \end{bmatrix}$$

where

$$a = 1 - \Delta tr - p\Delta t$$

$$b = \Delta tr$$

$$c = 1 - 2\Delta tr - p\Delta t$$

In this case the xenon and iodine dynamics can be written as

$$I_{j,k+1} = I_{j,k} - \Delta t\lambda_I I_{j,k} + \Delta t\gamma_I \Sigma_f \phi_{j,k}$$

$$X_{j,k+1} = X_{j,k} + \Delta t\lambda_I I_{j,k} + \Delta t\lambda_X \Sigma_f \phi_{j,k} - \Delta t\lambda_X X_{j,k} - \Delta t\sigma_X \phi_{j,k}$$

for $j = 0, 1, 2, \ldots, m$ and $k = 0, 1, 2, \ldots, n$.

4.8.6. Some Examples for Optimal Control Computations

4.8.6.1. Regulator Problem

The discrete version of equation (4.63) can be written as

$$J = \tfrac{1}{2}\Delta y \int_{0}^{t_f} [\Phi^{T}(t)Q\Phi(t) + U^{T}(t)RU(t)]\, dt$$

The lumped continuous maximum principle may be used to compute the optimal control, and the application of this maximum principle yields

$$\dot{\lambda}(t) = -[\Delta y Q \Phi(t) + A^T \lambda(t)], \qquad \lambda(t_f) = 0 \qquad (4.77)$$

$$U(t) = -\frac{1}{\Delta y} R^{-1} B^T \lambda(t)$$

The optimal control $u(y, t)$ can be solved from this two-point boundary value problem, if we assume that

$$\lambda(t) = P(t)\Phi(t) \qquad (4.78)$$

Substituting equation (4.78) into equations (4.73) and (4.77), we get

$$\dot{P}(t) = P(t)A - \frac{1}{\Delta y} P(t)BR^{-1}B^T P(t) + \Delta y QI + A^T P(t), \qquad P(t_f) = 0$$

$$\dot{\Phi}(t) = \left[A - \frac{1}{\Delta y} BR^{-1}B^T P(t) \right] \Phi(T), \qquad \Phi(0) = \Phi_0$$

$$U(t) = -\frac{1}{\Delta y} R^{-1} B^T P(t) \Phi(t)$$

$$u^*(t) = U(t) = I(t) + X(t)$$

Also the following equations must hold:

$$\dot{I}(t) = -\lambda_I I(t) + \gamma_I \Sigma_f \Phi(t), \qquad I(0) = I_0 \qquad (4.79)$$

$$\dot{X}(t) = \lambda_I(t) + \gamma_X \Sigma_f \Phi(t) - \lambda_X X(t) - \sigma_X \Phi(t), \qquad X(0) = X_0 \qquad (4.80)$$

Figure 4.2 shows the optimal control scheme for the regulator problem.

4.8.6.2. Numerical Example for the Regulator Problem

The following numbers are normalized quantities. We assume that

$$
\begin{array}{ll}
D = 0.02 & p = 0.0836 \\
y_f = 2.5 & t_f = 0.4 \\
R = 1.0 & Q = 1.0 \\
I_0(y) = 0.0 & X_0(y) = 0.0 \\
\lambda_I = 5 \times 10^{-9} & \lambda_I \Sigma_f = 9 \times 10^{-4} \\
\lambda_x = 7 \times 10^{-9} & \gamma_x \Sigma_f = 7 \times 10^{-4} \\
\sigma_X = 5 \times 10^{-4}
\end{array}
$$

and that

$$\Phi_0(y) = 1 + \alpha y, \qquad \alpha = 1.0$$

The optimal Φ and U are shown in Tables 4.1 and 4.2.

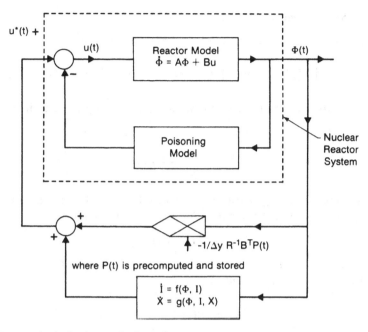

Figure 4.2. Optimal control scheme for nuclear reactor system regulator problem.

Table 4.1. Optimal Flux Variables (Φ) at Different Grid Points

			y			
t	0.0	0.5	1.0	1.5	2.0	2.5
0.0	1.0	1.5	2.0	2.5	3.0	3.5
0.1	0.96740	1.4458	1.9277	2.4096	2.8916	3.3700
0.2	0.94447	1.4062	1.8749	2.3436	2.8123	3.2741
0.3	0.93097	1.3808	1.8410	2.3013	2.7614	3.2113
0.4	0.92678	1.3694	1.8257	2.2821	2.7383	3.1810

Table 4.2. Optimal Control Variables ($-u$) at Different Grid Points

			y			
t	0.0	0.5	1.0	1.5	2.0	2.5
0.0	0.37355	0.55229	0.73625	0.92031	1.1042	1.2830
0.1	0.27816	0.41115	0.54811	0.68514	0.82210	0.95510
0.2	0.18476	0.27306	0.36402	0.45503	0.54599	0.63429
0.3	0.09237	0.13650	0.18197	0.22746	0.27293	0.31706
0.4	0.0	0.0	0.0	0.0	0.0	0.0

4.8.6.3. Servomechanism Problem

The discrete version of equation (4.67) can be written as

$$J = \tfrac{1}{2}\Delta y \int_0^{t_f} [\{\Phi(t) - \Psi(t)\}^T Q \{\Phi(t) - \Psi(t)\} + U^T(t) R U(t)] \, dt \quad (4.81)$$

The application of the lumped continuous maximum principle yields

$$\dot{\lambda}(T) = -\Delta y Q [\Phi(t) - \Psi(T)] - A^T \lambda(t), \qquad \lambda(t_f) = 0 \quad (4.82)$$

$$U(t) = -\frac{1}{\Delta y} R^{-1} B^T \lambda(t) \quad (4.83)$$

The optimal control $u(y, t)$ can be solved for from this two-point boundary value problem. To obtain a closed-loop control, we assume that

$$\lambda(t) = \eta(t) + P(t)\Phi(t) \quad (4.84)$$

where $\eta(t)$ and $P(t)$ are to be precomputed and stored. Using equation (4.84) in equations (4.82) and (4.73), we get

$$\dot{P}(T) + P(t)A - \frac{1}{\Delta y} P(t)BR^{-1}B^T P(t) + \Delta y Q I + A^T P(t) = 0 \qquad P(t_f) = 0$$

$$\dot{\eta}(T) - \frac{1}{\Delta y} P(t)BR^{-1}B^T \eta - \Delta y Q \psi + A^T \eta = 0 \qquad \eta(t_f) = 0$$

$$U(t) = -\frac{1}{\Delta y} R^{-1}B^T(\eta + P\Phi)$$

$$u^*(t) = U(t) = I(t) + X(t)$$

Also, equations (4.79) and (4.80) must hold. Figure 4.3 describes the optimal control scheme for the servomechanism problem.

4.8.6.4. Numerical Example for the Servomechanism Problem

We assume that

$D = 0.02$
$p = 0.0836$
$\phi_0(y) = 0.5 + y$ 　　　　 $\psi(y, t) = (1 + \beta y)(1 + \xi)$ 　 where $\beta = 1, \xi = 1.0$
$y_f = 2.5$ 　　　　　　　　　 $t_f = 0.4$
$R = 1$ 　　　　　　　　　　 $Q = 1$
$I_0(y) = 0.0$ 　　　　　　　 $X_0(y) = 0.0$

The other constants are as given in Section 4.8.6.2. The optimal ϕ and u are given in Tables 4.3 and 4.4.

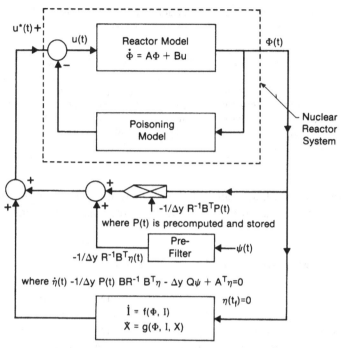

Figure 4.3. Optimal control scheme for the nuclear reactor system servomechanism problem.

4.8.7. Concluding Remarks

Most of the nuclear reactor systems are so widely distributed in space that approximate lumped model systems cannot be accurate. The purpose of this section was to extend the distributed optimal control computational techniques to a class of nuclear reactor systems. In the examples considered here, normalized states (neutron, flux, iodine, xenon), controls, and nuclear reactor parameters have been assumed. Although the examples taken here

Table 4.3. Optimal Flux Variables (ϕ) at Different Grid Points

	y					
t	0.0	0.5	1.0	1.5	2.0	2.5
0.0	0.5	1.0	1.5	2.0	2.5	3.0
0.1	0.48546	0.96388	1.45581	1.92775	2.40968	2.88810
0.2	0.47573	0.93751	1.40620	1.87494	2.34363	2.80541
0.3	0.47071	0.92063	1.38078	1.84104	2.30119	2.75111
0.4	0.47035	0.91306	1.36928	1.82571	2.28194	2.72465

Table 4.4. Optimal Control Variables (u) at Different Grid Points

				y		
t	0.0	0.5	1.0	1.5	2.0	2.5
0.0	0.22912	0.25480	0.27848	0.30208	0.32576	0.35144
0.1	0.17912	0.20349	0.22572	0.24789	0.27012	0.29448
0.2	0.12537	0.14540	0.16358	0.18173	0.19991	0.21994
0.3	0.06623	0.07835	0.08934	0.10030	0.11128	0.12340
0.4	0.0	0.0	0.0	0.0	0.0	0.0

are linear, the methods can be applied to nonlinear distributed nuclear reactor systems if suitable computational algorithms are devised.

4.9. Singular Perturbation Theory (Ref. 4.13)

In this section, we discuss a new technique used in 1976, based on singular perturbation theory, to obtain a suboptimal control law of nuclear reactors with spatially distributed parameters. The inverse of the neutron velocity is regarded as a small perturbing parameter. The model used has an infinite slab reactor described by the one-group diffusion equation. The importance of such a treatment has been increasing, since reactor sizes are increasing and spatial effects cannot be ignored.

The method in this section is based on the boundary layer method connected with a modal expansion analysis utilizing the Helmholtz mode. Because the inverse of the neutron velocity is a small quantity, the method enables us to construct an asymptotic series solution of each model coefficient, which is uniformly valid in the time interval considered. This method reduces the number of independent variables, thereby saving computation time and memory, and introduces time separation, which enables us to avoid the stiffness of the Riccati and trajectory equations.

We adopt the Helmholtz mode and show that the Helmholtz mode expansion is capable of transforming the state equation described by a set of partial differential equations of a singular perturbation type into an infinite set of ordinary differential equations of a singular perturbation type. The Kaplan mode, for instance, yields the ordinary differential equations, not of the singular perturbation type, which could in a broad sense be said to be of "singular" perturbation type, but of an awkward form.

Note also that we need a specific criterion relative to the criticality condition when reactors to be controlled are characterized by distributed parameters. Therefore, we show a difference of controllability, via the singular perturbation technique, that arises according to whether delayed neutrons are included.

4.9.1. State Equation

For simplicity, we consider an infinite slab reactor described by the one-group diffusion equation with one delayed neutron group as given by

$$\frac{1}{v}\frac{\partial}{\partial t}\phi(y, t) = \frac{\partial}{\partial y}\left[D(y)\frac{\partial}{\partial y}\phi(y, t)\right] - \Sigma_a(y)\phi(y, t) + (1 - \beta)\nu\Sigma_f(y)\phi(y, t)$$

$$+ \lambda C(y, t) - \Sigma_c(y, t)\phi(y, t) \tag{4.85}$$

$$\frac{\partial}{\partial t}C(y, t) = \beta\nu\Sigma_f(y)\phi(y, t) - \lambda C(y, t) \tag{4.86}$$

where

y = spatial coordinate
$\phi(y, t)$ = neutron flux
$C(y, t)$ = precursor density
$\Sigma_c(y, t)$ = absorption cross section, which is used for a control variable

The other notations have their usual meanings. The system described by equations (4.85) and (4.86) is called a *bilinear system*, and the linearization for this system is obtained by expanding each variable, of the distributed control variables and neutron flux, about the initial steady state.

The spatially dependent coefficient $D(y)$ prevents us from constructing analytical eigenfunctions. An approximation is made, following Iwasumi and Koga, avoiding this trouble so that the coefficient $D(y)$ is separated into two spatially variant and spatially invariant parts. In this method the other coefficients are also separated in the same way, ensuring the finality of the expansion series; the spatially independent parts should be chosen so that the fundamental mode can sustain its steady state and so that the effects of the spatially variant parts are taken into the control as the initial distribution of the control. This approximation gives satisfactory results if the system responds to the change; the validity depends on the problem considered. Here we consider a terminal cost problem in which the desired state is not far from the initial state.

A linearized state equation can be obtained by considering small deviations from the steady fundamental mode, which is determined by the system at the initial time, involving only spatially invariant parameters, and by normalizing with respect to the fundamental mode of the neutron flux ϕ_{10} as follows:

$$I_\varepsilon\frac{\partial}{\partial t}\phi(y, t) = S_y\phi(y, t) + Bu(y, t) \tag{4.87}$$

where

$$\phi(y, t) = \begin{bmatrix} \delta\phi(y, t)/\phi_{10} \\ \delta C(y, t)/\gamma_{10} \end{bmatrix}$$

$$u(y, t) = \frac{-\phi(y, t)\Sigma_c(y, t)}{\phi_{10}}$$

$$S_y = \begin{bmatrix} a_1 \partial^2/\partial y^2 + a_2 & a_3 \\ a_4 & a_5 \end{bmatrix}$$

$$I_\varepsilon = \begin{bmatrix} \varepsilon & 0 \\ 0 & 1 \end{bmatrix}, \qquad B = \begin{bmatrix} 1 \\ 0 \end{bmatrix}$$

and

$$a_1 = D, \qquad a_2 = [(1 - \beta)\nu\Sigma_f - \Sigma_a]$$

$$a_3 = \lambda, \qquad a_4 = \beta\nu\Sigma_f, \qquad a_5 = -\lambda, \qquad \varepsilon = \nu^{-1}$$

4.9.2. Problem Formulation

The problem for the nuclear reactor model described by the above equations can be stated as follows: obtain the optimal control that changes the neutron flux distribution from a given initial pattern to the desired state in a specified time T while minimizing the performance index given by

$$J = \int_0^l [\phi(y, T) - \phi_d(y)]^T Q[(y, t) - \phi_d(y)] \, dy$$

$$+ \int_0^T \int_0^l r[u(y, t) - u_0(y)]^2 \, dy \, dt$$

where

$$Q = \begin{bmatrix} \varepsilon q_1 & 0 \\ 0 & q_3 \end{bmatrix}, \qquad q_1, q_3, r > 0 \tag{4.88}$$

$\phi_d(y)$ is the desired state, and $u_0(y)$ is the initial value of the control determined beforehand so that the spatial homogeneity of the system parameters holds.

Note that the regulator problem of linear systems with terminal cost is mathematically equivalent to the problem stated here in that the deviations from the equilibrium in the regulator problem correspond to those from the desired state in the above problem. This is the case when the desired state is not far from the initial state.

4.9.3. Modal Expansion

Any orthonormal eigenfunction sequence can be used to solve partial differential equations, if the existence of the solution has been proven. In one-dimensional space (i.e., an infinite slab reactor), the Helmholtz mode expansion is appropriate and is defined by

$$\frac{d^2}{dy^2}\phi + \lambda\phi = 0 \qquad \text{with } \phi(0) = 0, \qquad \phi(l) = 0$$

where l is an extrapolated thickness; in this case the eigenfunction ϕ_i and the eigenvalue λ_i are given by

$$\phi_i = \left(\frac{2}{l}\right)^{1/2} \sin(\lambda_i)^{1/2} y \quad \text{and} \quad \lambda_i = \left(\frac{i\pi}{l}\right)^2$$

The sequence $\{\phi_i\}$ constitutes an orthonormal set in L^2 and is given by

$$\int_0^l \phi_i(y)\phi_j(y)\,dy = \delta_{ij} \tag{4.89}$$

where δ_{ij} is Kronecker's delta.

The state vector ϕ can be expanded into an infinite series with respect to the eigenfunctions generated above:

$$\phi(y, t) = \sum_{i=1}^{\infty} \phi_i(t)\left[\left(\frac{2}{l}\right)^{1/2} \sin\left(\frac{i\pi}{l}\right)y\right]$$

$$= \sum_{i=1}^{\infty} \left[\begin{array}{c}\xi_i(t)\\\eta_i(t)\end{array}\right]\left[\left(\frac{2}{l}\right)^{1/2} \sin\left(\frac{i\pi}{l}\right)y\right] \tag{4.90}$$

The control rods are physically concentrated and usually all at the top of the effective core, which can be represented mathematically by the Dirac function. When there are many control rods, however, the spatially concentrated control can be represented approximately by the spatially distributed control, which is simpler to analyze; in fact, the spatially concentrated case can be treated analogously. We restrict ourselves to the spatially distributed case for simplicity and clarity of exposition.

The control term $u(y, t)$ can be expanded with respect to the eigenfunction of the Helmholtz mode, if the control $u(\cdot, t)$ is given in $L^2(\Omega)$, the space of functions square integrable on Ω:

$$u(y, t) = \sum_{i=1}^{\infty} u_i(t)\left[\left(\frac{2}{l}\right)^{1/2} \sin\left(\frac{i\pi}{l}\right)y\right] \tag{4.91}$$

The modal coefficients $u_i(t)$, $\xi_i(t)$, and $\eta_i(t)$ are functions of time and are unknown. To obtain a set of equations for these modal coefficients, we substitute equations (4.90) and (4.91) into equation (4.87), using the orthonormality principle given by equation (4.89). We obtain

$$\varepsilon \frac{d\xi_i(t)}{dt} = \left[-a_1 \left(\frac{i\pi}{l} \right)^2 + a_2 \right] \xi_i(t) + a_3 \eta_i(t) + u_i(t) \tag{4.92}$$

$$\frac{d\eta_i(t)}{dt} = a_4 \xi_i(t) + a_5 \eta_i(t) \tag{4.93}$$

Equations (4.92) and (4.93) are decoupled with each mode, which results from the approximation made beforehand about the spatial homogeneity of parameters.

To synthesize the optimal feedback control in the mode space, we must expand the performance index into modal form. Thus, the desired terminal distribution $\phi_d(y)$ and the initial control $u_0(y)$ are expanded by using the same spatial eigenfunctions as follows:

$$\phi_d(y) = \sum_{i=1}^{\infty} \phi_{di} \left(\frac{2}{l} \right)^{1/2} \sin \left(\frac{i\pi}{l} \right) y = \sum_{i=1}^{\infty} \begin{bmatrix} \xi_{di} \\ \eta_{di} \end{bmatrix} \left(\frac{2}{l} \right)^{1/2} \sin \left(\frac{i\pi}{l} \right) y$$

$$u_0(y) = \sum_{i=1}^{\infty} u_{0i} \left(\frac{2}{l} \right)^{1/2} \sin \left(\frac{i\pi}{l} \right) y$$

and the modal expansion for the performance index can be obtained as follows:

$$J = \sum_{i=1}^{\infty} \tilde{\phi}_i^{\mathsf{T}}(T) Q \tilde{\phi}_i(T) + r \int_0^T \sum_{i=1}^{\infty} \tilde{u}_i^2(t) \, dt$$

The performance index J_i for each model i is

$$J_i = \frac{1}{2} \tilde{\phi}_i^{\mathsf{T}}(T) Q \tilde{\phi}_i(T) + \frac{1}{2} r \int_0^T \tilde{u}_i^2(t) \, dt \tag{4.94}$$

where

$$\tilde{\phi}_i^T(T) = \phi_i(T) - \phi_{di}$$

$$\tilde{u}_i(t) = u_i(t) - u_{0i}$$

From the above equations, we conclude that the optimal control theory for lumped parameter systems can be applied to each mode, where the coupling between each mode does not occur in the total system that includes the performance index.

Now the state equations of the ith mode and the corresponding performance index are given by equations (4.92) to (4.94), respectively. If we apply well-known results of optimal control theory, we can obtain the control law in feedback form:

$$u_i^*(t) = -\frac{1}{r} B^{\mathsf{T}} [K_i(t) \phi_i(t) + g_i(t)] + u_{0i}$$

where $K_i(t)$ is the solution of the matrix Riccati differential equation of the ith mode, which is given by

$$\frac{d}{dt} K_i(t) = -K_i(t)A_i + A_i^T K_i(t) + r^{-1} K_i(t) BB^T K_i(t)$$

where

$$A_i = \begin{bmatrix} \alpha_i/\varepsilon & a_3/\varepsilon \\ a_4 & a_5 \end{bmatrix}$$

$$\alpha_i = -a_1 \left(\frac{i\pi}{l}\right)^2 + a_2$$

and $g_i(t)$ is the solution of the following linear differential equation associated with the Riccati equation:

$$\frac{d}{dt} g_i(t) = -[A_i - r^{-1} BB^T K_i(T)] g_i(t) - K_i^T Bu_{0i}$$

If we assume that the matrix $K_i(t)$ is partitioned as

$$K_i(t) = \begin{bmatrix} \bar{k}_1 & \bar{k}_2 \\ \bar{k}_2 & \bar{k}_3 \end{bmatrix} = \begin{bmatrix} \varepsilon\tilde{k}_1 & \varepsilon\tilde{k}_2 \\ \varepsilon\tilde{k}_2 & \tilde{k}_3 \end{bmatrix}$$

and g_i as

$$g_i = \begin{bmatrix} \bar{g}_1 \\ \bar{g}_2 \end{bmatrix} = \begin{bmatrix} \varepsilon\tilde{g}_1 \\ \tilde{g}_2 \end{bmatrix}$$

then the following are Riccati equations for each element:

$$\varepsilon\frac{d}{dt} \tilde{k}_1 = -2\alpha_i\tilde{k}_1 + r^{-1}\tilde{k}_1^2 - 2\varepsilon a_4\tilde{k}_2 \tag{4.95}$$

$$\varepsilon\frac{d}{dt} \tilde{k}_2 = -a_3\tilde{k}_1 - \alpha_i\tilde{k}_2 - a_4\tilde{k}_3 + r^{-1}\tilde{k}_1\tilde{k}_2 - \varepsilon a_5\tilde{k}_2 \tag{4.96}$$

$$\frac{d}{dt} \tilde{k}_3 = -2a_3\tilde{k}_2 - 2a_5\tilde{k}_3 + r^{-1}\tilde{k}_2^2 \tag{4.97}$$

and the following are linear equations associated with $\tilde{g}_i(t)$:

$$\varepsilon\frac{d}{dt} \tilde{g}_1 = -(\alpha_i - r^{-1}\tilde{k}_1)\tilde{g}_1 - a_4\tilde{g}_2 - \tilde{k}_1 u_{0i} \tag{4.98}$$

$$\frac{d}{dt} \tilde{g}_2 = -a_3\tilde{g}_1 - a_5\tilde{g}_2 + r^{-1}\tilde{k}_2\tilde{g}_1 - \tilde{k}_2 u_{0i} \tag{4.99}$$

Equations (4.95) to (4.105) are solved with terminal conditions

$$K(T) = Q \tag{4.100}$$

and

$$g(T) = -Q\phi_d \qquad (4.101)$$

Note that in equations (4.95) through (4.101) the subscript of each variable representing the modal order is omitted. In the next section we shall omit this subscript unless confusion might occur.

In rectangular, cylindrical, and spherical geometries, the Helmholtz mode expansion can also be easily obtained analytically. The clean reactor mode is equivalent mathematically to the Helmholtz mode and leads to a satisfactory result.

4.9.4. Criticality Conditions and Applicability of Singular Perturbation Theory

The validity of the singular perturbation technique as applied to spatially dependent systems of reactors should be carefully checked in light of the theory of partial differential equations. A basic result in the theory of partial differential equations states that the Cauchy problem

$$\frac{\partial}{\partial t}\psi = c_1 \frac{\partial^2}{\partial w^2}\psi + c_2\psi \qquad (4.102)$$

is well posed and asymptotically stable as t tends to infinity, under the condition

$$c_1 > 0, \qquad c_2 < 0 \qquad (4.103)$$

If this condition holds, the eigenvalues of the equation

$$\left(c_1\frac{\partial^2}{\partial\omega^2} + c_2\right)\psi = \lambda\psi \qquad (4.104)$$

constitute a monotonically decreasing sequence

$$0 > c_2 > \lambda_1 \geq \lambda_2 \geq \cdots \geq -\infty \qquad (4.105)$$

Let us then observe the following criticality condition that should be satisfied by equation (4.83):

$$a_2 a_5 - a_3 a_4 - a_1 a_5 \left(\frac{\pi}{T}\right)^2 = 0$$

For a_1, a_3, $a_4 > 0$, and $a_5 < 0$, this condition yields the inequality

$$-a_1\left(\frac{i\pi}{T}\right)^2 + a_2 < 0, \qquad i = 1, 2, \ldots$$

which ensures that the stretched system of every mode,

$$\frac{d\xi_i}{d\tau} = \left[-a_1\left(\frac{i\pi}{l}\right)^2 + a_2\right]\xi_i + a_3\eta_i(T)$$

obtained in equation (4.92) by the transformation $\tau = t/\varepsilon$, is asymptotically stable as $\tau \to \infty$. This fact is apparent from the condition of equation (4.103); in other words, in the notation used in equation (4.83),

$$a_1 > 0, \qquad a_2 < 0$$

which obviously holds in our case. Thus, we establish the asymptotic accuracy of the synthesized distributed control and the trajectory given in asymptotic series form, which is shown in the next section, providing a secure error bound.

Equation (4.83), including delayed neutrons, has acquired stability. The kinetic equation without delayed neutrons—for example,

$$\frac{1}{v} \frac{\partial}{\partial t} \phi(\omega, t) = \frac{\partial}{\partial \omega} D \frac{\partial}{\partial \omega} \phi(\omega, t) + \Sigma_t \phi(\omega, t) + bu(\omega, t)$$

where $\Sigma_t = \nu \Sigma_f - \Sigma_a$, reveals a distinct situation concerning the stability or mode controllability. The criticality condition of this case is

$$D\left(\frac{\pi}{l}\right)^2 - \Sigma_t = 0$$

in the Helmholtz mode. This should hold at the initial time, so we have $D > 0$ and $\Sigma_t > 0$, which violates condition (4.103). In fact, the eigenvalues of the system satisfy

$$0 = \lambda_1 > \lambda_2 > \cdots > -\infty$$

which shows that the asymptotic stability of the stretched system is not realized for the first modal coefficient of the neutron flux. Therefore, the technique of singular perturbations cannot be applied to the first-mode system. In other words, physically the power control should be treated in the original form without using the approximation of singular perturbations; the spatial control can be dealt with by using singular perturbation theory. (Note that the power control is dominated by the fundamental mode, the spatial control by the higher modes.) This differs from the case of delayed neutrons, in which both power and spatial controls can be achieved by singular perturbations.

Although mode decoupling enables us to apply singular perturbation theory to each mode system separately, such an argument as above is further required of the stability of system solutions before we apply the existing singular perturbation technique of ordinary differential equations.

4.9.5. Construction of Asymptotic Expansions

To find the solutions of equations (4.95) through (4.99) for small ε $(=v^{-1})$, we apply the boundary layer method developed for the lumped parameter systems. The solutions are in the form

$$\tilde{k}_j = \tilde{k}_j(t, \varepsilon) = k_j(t, \varepsilon) + h_j(\tau, G), \qquad j = 1, 2, 3, \ldots$$

where

$$\tau = \frac{T - t}{\varepsilon}$$

and $k_j(t, \varepsilon)$, $h_j(\tau, \varepsilon)$ admit asymptotic expansions in ε, and ε tends to zero in the form of

$$k_j(t, \varepsilon) = \sum_{r=0}^{\infty} k_j^r(t)(r!)^{-1}\varepsilon^r \tag{4.106}$$

$$h_j(\tau, \varepsilon) = \sum_{r=0}^{\infty} h_j^r(\tau)(r!)\varepsilon^r \tag{4.107}$$

The sequences of the right sides of equations (4.106) and (4.107) converge to the solutions of the Riccati integrodifferential equation and of the stretched Riccati equation associated with it, respectively.

Substituting the outer expansion

$$\sum_{r=0}^{\infty} k_j^r(r!)^{-1}\varepsilon^r \tag{4.108}$$

into equations (4.95)–(4.97) and comparing the coefficients of the same power of ε, we obtain the following set of equations. The first set correspond to $r = 0$, the second to any value of r:

$$0 = -2\alpha_i k_1^0 + r^{-1}(k_1^0)^2 \tag{4.109}$$

$$0 = -a_4 k_3^0 - \alpha_i k_2^0 - a_3 k_2^0 + r^{-1} k_1^0 k_2^0 \tag{4.110}$$

$$\frac{d}{dt} k_3^0 = -2a_5 k_3^0 - 2a_3 k_2^0 + r^{-1}(k_2^0)^2 \tag{4.111}$$

Also,

$$\frac{d}{dt} k_1^{r-1} = -2\alpha_i k_1^r + 2r^{-1} k_1^0 k_1^r + p_1^r(t) \tag{4.112}$$

$$\frac{d}{dt} k_2^{r-1} = -a_3 k_1^r - \alpha_i k_2^r - a_4 k_3^r + r^{-1}(k_1^0 k_2^r + k_1^r k_2^0) + p_2^r(t) \tag{4.113}$$

$$\frac{d}{dt} k_3^r = -2a_3 k_2^r - 2a_5 k_3^r + 2r^{-1} k_2^0 k_2^r + p_3^r(t) \tag{4.114}$$

where the remainders $p_j(t)$, $j = 1, 2, 3$, are polynomials involving the known terms in the preceding steps.

To construct a uniformly valid expansion, we consider the boundary layer corrections. The recursive set of equations satisfied by these corrections is obtained by using the stretched system

$$-\frac{d}{d\tau} h_1 = -2\alpha_i h_i + r^{-1}(h_1^2 + 2k_1 h_1) - 2\varepsilon a_4 h_2 \tag{4.115}$$

$$-\frac{d}{d\tau} h_2 = -\alpha_3 h_1 - a_i h_2 - a_4 h_3 - \varepsilon a_5 h_2 + r^{-1}(h_1 h_2 + k_2 h_1 + k_1 h_2) \tag{4.116}$$

$$-\frac{d}{d\tau} h_3 = \varepsilon[-2a_3 h_2 + r^{-1}(h_2^2 + 2k_2 h_2) + r^{-1} h_1^0 - 2a_5 h_3] \tag{4.117}$$

The resulting recursive set of equations thus becomes

$$-\frac{d}{d\tau} h_1^0 = -2\alpha_i h_1^0 + r^{-1} h_1^0 + r^{-1} h_1^0[h_1^0 + 2k_1^0(T)] \tag{4.118}$$

$$-\frac{d}{d\tau} h_2^0 = -a_3 h_1' - \alpha_i h_2^0 - a_4 h_3^0 + r^{-1}[h_1^0 h_2^0 + h_1^0 k_2^0(T) + h_2^0 k_1^0(T)] \tag{4.119}$$

$$-\frac{d}{d\tau} h_3^0 = 0 \tag{4.120}$$

For the higher order,

$$-\frac{d}{d\tau} h_1^r = -2\alpha_i h_1^r + r^{-1}[2h_1^0 h_1^r + 2h_1^r k_1^0(T)] + w_1^r(T) \tag{4.121}$$

$$-\frac{d}{d\tau} h_2^r = -a_3 h_1' - \alpha_i h_2^r - a_4 h_3^r + r^{-1}[h_1^0 h_2^r + h_1^r h_2^0 + h_1^r k_2^0(T) + h_2^r k_1^0(T)]$$

$$+ w_2^r(\tau) \tag{4.122}$$

$$-\frac{d}{d\tau} h_3^r = w_3^r(\tau) \tag{4.123}$$

where the remainders $w_i^r(\tau)$ consist only of functions decaying exponentially.

The terminal conditions for these recursive sets of equations (4.109) through (4.123) can be derived in the following way. For equations (4.109) to (4.111), the zeroth-order equation in the outer region, we let

$$k_3^0(t) = q_3 \tag{4.124}$$

Now, using equation (4.124) with equations (4.109) through (4.111), we can determine values for $k_1^0(T)$ and $k_2^0(T)$. Also, the complete expansion satisfying the original terminal condition is

$$k_j(T) + h_i(0) = q_j, \qquad j = 1, 2, 3, \ldots \tag{4.125}$$

where q_j are given in equation (4.88) ($q_2 = 0$) and are assumed to be independent of ε.

From equations (4.124) and (4.125) we can find the terminal conditions for equations (4.118) through (4.120) as

$$h_1^0(0) = q_1 - k_1^0(T) \tag{4.126}$$

$$h_2^0(0) = -k_2^0(T) \tag{4.127}$$

$$h_3^0(0) = 0 \tag{4.128}$$

Solutions to equations (4.118) through (4.120) with equations (4.126) through (4.128) lead to the zeroth-order coefficients $k_j^0(t)$ and $h_j^0(\tau)$.

The terminal conditions for the first-order coefficients $k_j^1(t)$ and $h_j^1(\tau)$ are determined as follows, the procedure being peculiar to singular perturbations. Since the remainder $w_3'(\tau)$ in equations (4.121)–(4.123) consists only of functions decaying exponentially as τ tends to infinity, the integration of equation (4.123) leads to

$$h_3^1(\tau) = \int_\tau^\infty w_3^1(\tau) \, d\tau$$

where

$$w_3^1(\tau) = -2a_3 h_2^0(\tau) - 2a_5 h_3^0(\tau) + r^{-1}\{[h_2^0(\tau)]^2 + 2k_2^0(T)h_2^0(\tau)\}$$

then, the terminal condition of the first boundary layer correction $h_3^1(0)$ can be determined by

$$h_3^1(0) = \int_0^\infty w_3^1(\tau) \, d\tau$$

The integrand involves only the terms determined in the zeroth-order boundary layer, with the terminal condition. Thus, the terminal condition of the first-order coefficient of the outer expansion $k_3^1(T)$ can be derived by using equation (4.12), resulting in

$$k_1^3(\tau) = -\int_0^\infty w_3^1(\tau) \, d\tau$$

under which the first correction system of the outer expansion can be solved thoroughly. Then we can evaluate each coefficient of the first outer expansion at the terminal time $t = T$. The values of the boundary layer correctors at the terminal time $\tau = 0$ directly follow by using equation (4.125):

$$h_1^1(0) = -k_1^1(T), \qquad h_2^1(0) = -k_2^1(T)$$

Thus, we can carry out the computation of the first boundary layer correctors given in equations (4.121)-(4.123).

The above procedure determines the total system of the reduced and first orders completed. A similar result can be derived for the higher-order systems by using the same algorithm. We can also apply a similar procedure to the linear differential equation for $\tilde{g}_j$, but we omit the details.

The practical computation prohibits treating infinite terms of the modal expansion, and hence the truncated series should be adopted at a finite Nth mode. The suboptimal distributed control is then

$$u_{\mathrm{sub}}(\omega, t) = \sum_{i=1}^{N} u_i^*(t) \left(\frac{2}{l}\right)^{1/2} \sin\left(\frac{i\pi}{l}\right)\omega$$

the suboptimal trajectory, spatially dependent, is

$$\Phi_{\mathrm{sub}}(\omega, t) = \sum_{i=1}^{N} \Phi_i(t) \left(\frac{2}{l}\right)^{1/2} \sin\left(\frac{i\pi}{l}\right)\omega$$

where the optimal trajectory Φ_i of the ith mode can be derived as a solution of the corresponding trajectory equation

$$\varepsilon \frac{d}{dt}\xi_i = (\alpha_i - r^{-1}\tilde{k}_{i1})\xi_i + (a_3 - r^{-1}\tilde{k}_{i2}\eta_i - r^{-1}\tilde{g}_{i1} + u_{0i}) \qquad (4.129)$$

$$\frac{d}{dt}\eta_i = a_4\xi_i + a_5\eta_i \qquad (4.130)$$

with initial conditions

$$\xi_i(0) = \left(-\alpha_i + \frac{a_3 a_4}{a_5}\right)^{-1} u_i(0)$$

$$\eta_i(0) = \frac{a_4}{a_5}\xi_i(0) \qquad (4.131)$$

which can be obtained by equating the differential terms in equations (4.92) and (4.93) to zero. The above procedure for computing the initial condition, equation (4.131), is reasonable from the requirement of the steady-state sustenance at the initial time. Equations (4.129) and (4.130) can be treated approximately by singular perturbation theory, as shown in the lumped parameter case.

In general, once we have derived the asymptotic expansion truncated at the $(n+1)$th terms, there exist functions $S_i^n(t, \varepsilon)$ and $R_j^n(t, \varepsilon)$ bounded uniformly in the interval $[0, T)$ such that

$$\tilde{k}_j(t) = \sum_{r=0}^{n} [k_j^r(t) + h_j^r(\tau)](r!)^{-1}\varepsilon^r + S_j^n(t, \varepsilon)\varepsilon^{n+1}$$

$$\tilde{g}_j(t) = \sum_{r=0}^{n} [g_j^r(t) + d_j^r(\tau)](r!)^{-1}e^r + R_j^n(t, \varepsilon)\varepsilon^{n+1}$$

for each mode. A similar result applies to the neutron trajectory and the control.

The error due to truncation at the $(n + 1)$st term is of the order of ε^{n+1} in the modal expanded space. Thus, if the required number of modes is adopted, we can derive the overall approximate solution with errors due to truncation of the asymptotic series solution in ε. The model coefficient $K_i(t)$ is equivalent to the kernel generated from the operator satisfying the Riccati equation in an operator sense.

4.9.6. Practical Numerical Example

Reference 4.13 tried to compute numerically approximate solutions belonging to the first to the fifth modes to compare with the corresponding exact solutions for a system exemplified by data given in Table 4.5, where $\varepsilon = \nu^{-1} = 0.45 \times 10^{-5}$ sec/cm. The approximate solutions have been derived in the form of series truncated at the second term. The results obtained for the first and second modes are shown in Figures 4.3, 4.4, and 4.5.

Usually, the feedback coefficients of each mode change rapidly in the neighborhood of the terminal time, leading to computational difficulties that become much more serious than in the lumped parameter case, since equations of many modes must be treated and the existence of the boundary layers appearing in every mode makes the change more rapid. The method presented in this chapter can reduce such difficulties. Figure 4.3 shows the time behaviors of feedback coefficients $\bar{k}_j$. The approximation to each feedback coefficient has good accuracy as expected except for $\bar{k}_1$, which happens to be relatively small. The error of $\bar{k}_1$ of the first and second modes is about 10^{-12} and 10^{-14}, respectively, which is consistent with the theoretical evaluation indicated at the end of the preceding section.

Table 4.5. Data Used in the Example

λ	0.078 sec^{-1}	1	100 cm
Σ_f	0.0202 cm^{-1}	T	10 sec
Σ_a	0.05 cm^{-1}	r	1000
D	0.5066 cm	q_1	1
ν	2.2×10^5 cm/sec	q_3	78

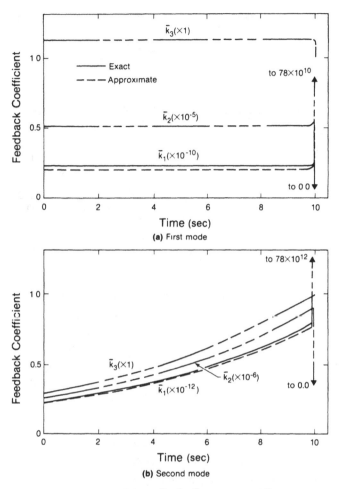

Figure 4.4. Feedback coefficients $\bar{k}_1$, $\bar{k}_2$, $\bar{k}_3$, elements of $\bar{k}_i$, $i = 1, \ldots, 3$.

Figure 4.4 shows the time behaviors of $\bar{g}_j$ whose approximate solutions give satisfactory results. The error is about 0.1%. Figure 4.5 shows the time behaviors of amplitudes of the neutron flux.

In the third to fifth modes, qualitative tendencies similar to those above are observed, with rapid change near the terminal time for all feedback coefficients $\bar{k}_j$, $\bar{g}_j$, and amplitudes of the neutron flux, with the exception of the fourth mode, where $\bar{g}_j$ and the amplitudes of the neutron flux remain almost zero throughout the entire control period.

Figure 4.6 shows the exact and approximate transients of the neutron flux shape in this example, both being obtained by using the first to the fifth modes.

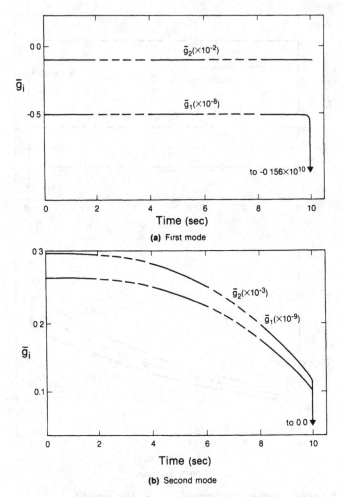

Figure 4.5. Values of $\bar{g}_1$, $\bar{g}_2$, elements of the vector g_i, $i = 1, 2$.

A numerical study shows that the computation time can be reduced to 37% of that in the case of the exact solution (the original modal solution). This saving emphasizes the fact that the method presented is very efficient in various points.

4.9.7. Concluding Remarks

An approximate solution of an optimal problem, with terminal cost, arising in distributed parameter nuclear reactors was derived. The method is based on singular perturbation theory, which gives the precise error estimation with respect to the truncation of the asymptotic series solution

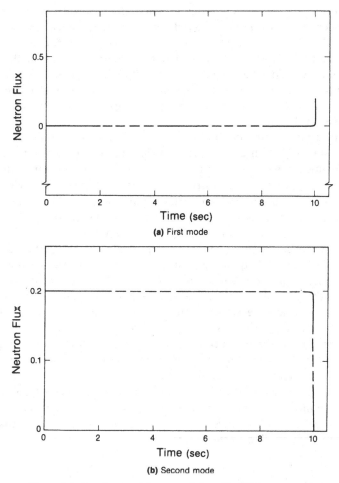

Figure 4.6. Amplitude of the neutron flux $\bar{\xi}_i(2/l)^{1/2}\xi_i$, $i = 1, 2$.

and leads to a solution uniformly valid in the time interval considered. The efficiency of the method has been ascertained by showing a numerical study of an infinite slab reactor.

It has been shown that when one employs the Helmholtz mode, as in this chapter, the mode expansion yields ordinary differential equations of an ideal singular perturbation type. The choice of mode should be made carefully: the Kaplan mode appears to be inadequate to apply to the singular perturbation theory.

It has also been clarified that mode controllability via the singular perturbation technique depends on the criticality condition. In this connection, a basic theorem of partial differential equations has been shown to play an important role. Reference 4.13 actually indicated a difference that arises according to whether delayed neutrons are included: for the former case, both power and spatial controls are possible, but for the latter only spatial control is possible. Note that the criterion based on the partial differential equations theory is more substantial than that based on observing differential equations obtained through the mode expansion. Furthermore, a complement is added below on another possible method utilizing singular perturbation theory.

We summarize the method adopted in the previous sections. First, the original state equation described by partial differential equations is expanded into modal form, resulting in an infinite set of ordinary differential equations. Then singular perturbation theory is applied to it, yielding an infinite recursive set of ordinary differential equations. The solution gives coefficients of asymptotic expansions of the optimal control and trajectory of each mode. On the other hand, there is another way: we can apply singular perturbation theory directly to the original state equation described by partial differential equations, thus generating an infinite recursive set of partial differential equations whose number of independent variables is reduced; the generated recursive set can then be expanded into modal form, resulting in an infinite set of ordinary differential equations within each mode that can be properly solved.

These two methods are equivalent and give solutions converging to the exact solution satisfying the original control problem. The convergence is rigorously proved by Asatani (Ref. 4.13).

In the second step of the latter procedure, it is suggested that methods other than modal expansion be used, such as the nodal expansion method or the direct finite-difference method, for example. It is hoped that these procedures are studied in the future.

4.10. System Tau Method (Ref. 4.15)

In this section, we discuss another efficient method for solving systems of coupled ordinary differential equations with initial, boundary, and/or intermediate conditions given; the *system tau method* (STM). Optimal control problems of nuclear systems can be solved by this technique, in which Pontryagin's maximum principle is used to transform the state equations and their associated performance index into a system of coupled differential equations with mixed boundary conditions. Hence, the solution is found in a noniterative manner, and any nonlinear equations are linearized, if it is valid to do so.

At present, STM is limited to systems of linear ordinary differential equations whose coefficients have polynomial form. The systems may have any combination of given initial, boundary, and intermediate conditions. Realistic optimal control problems may be put into this form by changing the state equations from a system of partial differential equations to a system of ordinary differential equations by using some variable reduction scheme, such as the modal or nodal techniques; we then apply Pontryagin's maximum principle to obtain a set of coupled ordinary differential equations with mixed boundary conditions. In the next section, we discuss this technique in detail, and compare it with others.

4.10.1. Calculational Method

The method described here differs from the usual solution approach, which attempts to approximate the solution of the exact problem. Instead, this method finds the exact solution to an approximate problem through an expansion technique. The concept of an expansion seems to be desirable due to its straightforward solution method. However, the question of what the expansion functions should be remains. Rather than guessing or relying on a priori knowledge to find the functions, as is done in synthetic solutions, this method calculates the expansion functions as part of the solution. It is an extension to systems of equations of the method developed by Lanczos (Ref. 4.16) for a single equation. The solution technique leads to a system of algebraic equations that determine the expansion functions in polynomial form. The final solution is found by using the given boundary conditions. We assume that this solution is sufficiently smooth to be defined by polynomials, has no more zeros than the order of polynomials, and is not singular.

Can this technique be applied to many realistic physical problems? The solutions to these problems are continuous; however, their derivatives may not be. Therefore, since polynomials are continuous in nature and arbitrarily differentiable, can they approximate a function that is continuous but may have a discontinuity in one of its derivatives? The Weierstrass approximation theorem answers this question.

Theorem 4.1 (Weierstrass Approximation Theorem). If $f(x)$ is continuous on a finite interval $[a, b]$, then, given any $\varepsilon > 0$, there exists an $n = n(\varepsilon)$ and a polynomial $p_n(x)$ of degree n such that $|f(x) - P_n(x)| < \varepsilon$ for all x in $[a, b]$.

Using Weierstrass's theorem, we can approximate the solutions to physical problems that may have discontinuous derivatives, using expansion by polynomial functions.

The solution method is described for the class of problems consisting of a system of inhomogeneous equations with constant coefficients; i.e.,

$$\dot{X}_i(t) - \sum_{k=1}^{n} P_{ik} X_k(t) = f_i(t), \qquad i = 1, \ldots, n, \qquad t > 0 \quad (4.132)$$

The jth-order approximation for equation (4.132) (i.e., where the functions x_i and f_i are expanded into jth-order polynomials and an error term is added) is

$$\dot{X}_{ij}(t) - \sum_{k=1}^{n} p_{ik} X_{kj}(t) = f_{ij}(t) + \left(\tau_i - \frac{d_{i0}}{C_j^0} \right) \sum_{m=0}^{j} C_j^m t^m, \qquad i = 1, \ldots, n$$

(4.133)

where

$$f_i(t) = \sum_{m=0}^{N_i} d_{im} t^m, \qquad i = 1, \ldots, n$$

and its jth approximation is

$$f_{ij}(t) = \sum_{m=0}^{i} d_{im} t^m, \qquad i = 1, \ldots, n \qquad (4.134)$$

with $d_{im} = 0$ for $m > N_i$. This means that for $j < N_i$ some terms are lost in the approximate problem. Therefore, when $j < N_i$, the approximation to the actual problem is less accurate due to the neglected terms. When the inhomogeneity is a function that has an infinite power series expansion, the inhomogeneity should be removed by the addition of new independent variables and their appropriate equations to the system. For example, if $f(t) = be^{at}$, then the additional equation should be

$$\dot{X}_{n+1}(t) = aX_{n+1}(t), \qquad X_{n+1}(0) = b$$

Although this procedure slightly increases the computational effort per step, experience has shown that the accuracy of the solution is greatly improved with lower orders of approximation.

The additional term on the right side of equation (4.133),

$$\left(\tau_i - \frac{d_{i0}}{C_j^0} \right) \sum_{m=0}^{j} C_j^m t^m = \left(\tau_i - \frac{d_{i0}}{C_j^0} \right) T_j^*(t)$$

is the error term added to get an approximate problem that can be solved exactly by a jth-order polynomial. The τ_i are unknown constants that are determined by the given mixed boundary conditions on the exact problem. The variable $T_j^*(t)$ represents a set of orthogonal polynomials, such as the shifted Chebyshev polynomials. In this section the Chebyshev polynomials are used because they minimize the maximum error over a region, and this is frequently the desired behavior for physical problems. However, any set of orthogonal polynomials may be used, and should be, if it exhibits some properties advantageous to the problem. The only difference in finding the solution caused by using different polynomials is in determining which set of tabulated coefficients C_j^m is used.

We assume that the solution to equation (4.133) is

$$X_{ij}(t) = \sum_{m=0}^{j} C_j^m Q_{im}(t) \tag{4.135}$$

where the $Q_{im}(t)$ are "canonical" polynomials. This solution differs in a basic manner from the type Lanczos used for the single-equation case. There the error multiplier τ_i was a constant multiplier of the weighted canonical polynomials defining the solution. These polynomials were a function of t only. In the system case, the τ_k, $k = 1, \ldots, n$, are absorbed in the canonical polynomials themselves so that Q_{im} is a function of t and τ_k; however, the dependence on τ_k is not written explicitly for the sake of simplicity, but it should be kept in mind. The additional term d_{i0}/C_j^0 is separated from the τ_i to allow easy separation of the homogeneous solution and the particular solution.

Through the use of equations (4.134) and (4.135), equation (4.133) becomes

$$\sum_{m=0}^{j} C_j^m \dot{Q}_{im}(t) - \sum_{k=1}^{n} P_{ik} \sum_{m=0}^{j} C_m^J Q_{km}(t) = \sum_{m=0}^{j} d_{im} t^m + \left(\tau_i \frac{d_{i0}}{C_j^0} \right) \sum_{m=0}^{j} C_j^m t^m$$

$$i = 1, \ldots, n \tag{4.136}$$

For a given m, if the Q_{km} for $k = 1, \ldots, n$ satisfy

$$\dot{Q}_{im}(t) - \sum_{k=1}^{n} P_{ik} Q_{km}(t) = \left(\tau_i + \frac{d_{im}}{C_j^m} - \frac{d_{i0}}{C^0} \right) t^m, \qquad i = 1, \ldots, n \tag{4.137}$$

they will satisfy equation (4.136). This is not the only set of equations defining the $Q_{km}(t)$ that would satisfy equation (4.136), but it is a convenient one.

The desired form of the canonical polynomials is

$$Q_{km}(t) = \sum_{r=0}^{m} a_{kmr} t^r \tag{4.138}$$

where the a_{kmr} are constants to be determined.

Using equation (4.138) in equation (4.137) yields

$$\sum_{r=0}^{m-1} (r+1) a_{im,r+1} t^r - \sum_{k=1}^{n} P_{ik} \sum_{r=0}^{m} a_{kmr} t^r = \left(\tau_i + \frac{d_{im}}{C_j^m} - \frac{d_{i0}}{C_j^0} \right) t^m,$$

$$i = 1, \ldots, n \tag{4.139}$$

By equating like powers of t in equation (4.139), we obtain the recurrence relations

$$\sum_{k=1}^{n} P_{ik}a_{kmr} = (r+1)a_{im,r+1}, \quad r = 0, \ldots, m-1 \qquad i = 1, \ldots, n$$

$$(4.140)$$

$$\sum_{k=1}^{n} P_{ik}a_{kmm} = -\tau_i + \frac{d_{i0}}{C_j^0} - \frac{d_{im}}{C_j^m}, \quad r = m \qquad i = 1, \ldots, n$$

It is obvious that there is only one $n \times n$ matrix (i.e., $[P_{ik}]$) to be inverted to obtain a_{kmr}.

Suppose that d_{ik}, $k = 0, \ldots, j$, are all zero. Then three important computational properties of the canonical polynomials are obvious. First, although the theory required to obtain the relations in equation (4.140) is involved, the recurrence relations can be written directly from the original problem. Second, $Q_{km}(t)$ are independent of the order of approximation j. This means that, if the jth approximation to the solution has been found but is not sufficiently accurate, higher-order approximations use the same first $j + 1$ canonical polynomials and multiply them by a different set of tabulated coefficients so that the original work is not wasted. Third, it can be seen from the form of equation (4.140) that $a_{k,m+1,r+1} = [(m+1)/(r+1)]a_{kmr}$. This is important computationally because if the mth canonical polynomial is known, then the $(m + 1)$ canonical polynomial is also known, with the exception of the coefficient of t^0. The n coefficients of t^0 are found by using the inverse of $[P_{ik}]$, so only one matrix needs to be inverted no matter how high the order of approximation.

If the d_{ik}, $k = 0, \ldots, j$, are not all zero, it appears that the second advantage mentioned is lost. This apparent dependence of the canonical polynomials on the order of the approximation used can be eliminated. The solution of an inhomogeneous problem is the solution of the homogeneous problem plus the particular solution due to the terms d_{i0}/C_j^0 and d_{im}/C_j^m. Therefore, a_{kmr} can be broken up into three terms—$a_{kmr} = a_{kmr}^{[1]} + a_{kmr}^{[2]} + a_{kmr}^{[3]}$—so that equation (4.140) becomes

$$\sum_{k=1}^{n} P_{ik}a_{kmr}^{[s]} = (r+1)a_{im,r+1}^{[s]}, \quad s = 1, 2, 3 \quad r = 0, \ldots, m-1$$

$$= -\tau_i, \quad s = 1 \qquad i = 1, \ldots, n$$

$$(4.141)$$

$$\sum_{k=1}^{n} P_{ik}a_{kmm}^{[s]} = \frac{d_{i0}}{C_j^0}, \quad s = 2 \qquad r = m$$

$$= -\frac{d_{im}}{C_j^m}, \quad s = 3 \qquad i = 1, \ldots, n$$

where $a_{kmr}^{[1]}$ = coefficients of the homogeneous problems

$a_{kmr}^{[2]}$ = coefficients due to the additional error term relating to the t^0 term of the inhomogeneity

$a_{kmr}^{[3]}$ = coefficients due to the t^m term of the inhomogeneity

The $a_{kmr}^{[1]}$ are independent of C_j^0 and C_j^m and, therefore, are independent of the order of approximation. The terms $1/C_j^0$ and $1/C_j^m$ are constants and, since $Acx = cAx$, where c is a constant and A is a matrix, the quantities $\bar{a}_{kmr}^{[2]}$ and $\bar{a}_{kmr}^{[3]}$ may be calculated from equation (4.141) by neglecting C_j^0 and C_j^m. Therefore they are independent of the order of approximation, and so is the canonical polynomial. When the order of approximation q is decided on, the actual coefficients may be calculated from

$$a_{kmr}^{[2]} = \frac{1}{C_q^0 \bar{a}_{kmr}^{[2]}} \quad \text{and} \quad a_{kmr}^{[3]} = \frac{1}{C_q^m \bar{a}_{kmr}^{[3]}}$$

To complete the problem, we obtain the τ_i values from the given conditions (initial, boundary, and/or intermediate).

To illustrate the method, consider the following problem, which has an analytical solution:

$$\dot{X}_1(t) + X_1(t) - \tfrac{1}{2}X_2(t) = 1 + t^2, \qquad X_1(0) = 2$$

and

$$\dot{X}_2(t) - 2X_1(t) - X_1(t) - X_2(t) = t, \qquad X_2(1) = 0$$

where $0 \leq t \leq 1$. The jth approximate problem from equation (4.133) is

$$\dot{X}_{1j}(t) + X_{1j}(t) - \tfrac{1}{2}X_{2j}(t) = 1 + t^2 + \left[\tau_1 - \frac{1}{C_j^0}\right] \sum_{m=0}^{j} C_j^m t^m$$

and

$$\dot{X}_{2j}(T) - 2X_{1j}(t) - X_{2j}(T) = t + \tau_2 \sum_{m=0}^{j} C_j^m t^m$$

From equation (4.135),

$$X_{1j}(t) = \sum_{m=0}^{J} C_j^m Q_{1m}(t)$$

and

$$X_{2j}(t) = \sum_{m=0}^{j} C_j^m Q_{2m}(t)$$

where, according to equation (4.138),

$$Q_{1m}(t) = \sum_{m=0}^{J} a_{1mr}(t)$$

and

$$Q_{2m}(t) = \sum_{r=0}^{m} a_{2mr}(t)$$

The recurrence relations are

$$-a_{1mr} + \tfrac{1}{2}a_{2mr} = (r+1)a_{1m,r+1},$$

$$a_{21mr} + a_{2mr} = (r+1)a_{2m,r+1}, \qquad r = 0, \ldots, m-1$$

and

$$-a_{1mm} + \tfrac{1}{2}a_{2mm} = -\tau_1 + \frac{1}{C_j^0} - \frac{d_{im}}{C_j^m},$$

$$a_{1mm} + a_{2mm} = -\tau_2 - \frac{d_{2m}}{C_j^m}, \qquad r = m$$

The first two canonical polynomials for each variable are

$$Q_{10}(t) = \frac{1}{2}\tau_1 - \frac{1}{4}\tau_2$$

$$Q_{20}(t) = -\tau_1 - \frac{1}{2}\tau_2$$

$$Q_{11}(t) = \frac{1}{2}\tau_1 + \left(\frac{1}{2}\tau_1 - \frac{1}{4}\tau_2\right)t + \frac{1}{C_j^0}\left(\frac{1}{2} - \frac{1}{2}t\right) = \frac{1}{4}\frac{1}{C_j^1}$$

and

$$Q_{21}(t) = -\frac{1}{2}\tau_2\left(\tau_1 + \frac{1}{2}\tau_2\right)t + \frac{1}{C_j^0} - \left(\frac{1}{2} + \frac{1}{2}t\right)\frac{1}{C_j^1}$$

The given mixed boundary conditions uniquely determine τ_1, τ_2 and, hence, $X_1(t)$ and $X_2(t)$. The above calculation, although difficult by hand, is trivial on the computer. The fifth-order approximation for this problem is good to 5 significant digits and the tenth is good to 12 digits.

Exclusive of any computer time and space arguments, this method has the following inherent advantageous properties.

1. Due to the manner in which the method satisfies given conditions, it handles initial conditions, end-point conditions, intermediate-point conditions, and any combination of these with equal ease. This means that the time and effort necessary to solve any of these types of problems are the same as for solving the initial value problem.

2. For any system of state equations, once the canonical polynomials have been calculated for that system, parametric studies may be done where the given conditions are varied without having to solve the problem each time. The only thing to be done for each case is to reevaluate the τ-values. This can be important in a reactor control problem in which it is desired to find the optimal control for changes between various initial and final power levels.

3. For the inhomogeneous problem, once the homogeneous canonical polynomials have been found, a parametric study of different forcing functions may be made by solving only for the inhomogeneous part of the canonical polynomials each time. This can be a great savings over having to solve the entire problem each time.

4. Once the canonical polynomials have been obtained for a system of state equations over a given range of the independent variable, problems over different ranges can be solved with the same canonical polynomials. In this situation, the only change is a constant multiplier for the tabulated coefficients. This means that if the same system is applied to sets of problems over many time ranges, the problem need not be solved again each time except for the τ-values. Problems over different regions may also be solved by rescaling the problem.

5. When the method uses Chebyshev polynomials, the maximum error is minimized. However, if some other orthogonal polynomial's trait is desired, the only difference is the set of tabulated coefficients that is used. If a study of how some different orthogonal polynomials affect the solution is desired, the canonical polynomials need be calculated only once.

6. The methods takes advantage of "flexible" coefficients to improve the accuracy of the solution by using fewer terms than with "rigid" coefficients. This is done without paying the price of recalculating all of the coefficients when a higher-order approximation is desired. It is accomplished by weighting rigid canonical polynomials with different tabulated coefficients.

7. The method is not iterative.

8. The solution is found as an explicit expression. This means that the solution can be found at any point in the range without having to redo the entire problem if that value was not originally evaluated.

The method used for finding bounds on the error consists basically of finding a bound on $|y(x) - y_n(x)|$, where $y(x)$ is the solution to the exact problem and $y_n(x)$ is the solution to the approximate problem. Since the desired relationship may be found for systems of equations with initial values, the approach is to find an equivalent initial value problem to the mixed boundary problem and to derive the inequality for that problem. Through relationships between the solution of the initial value problems and the mixed boundary value problems, the expression for the inequality is transformed to one for the mixed boundary value problem. The error bounds calculated for the above inhomogeneous problem are 3×10^{-2} for the fifth-order approximation and 8×10^{-9} for the tenth-order one. In this case, the bounds are clearly very conservative.

4.10.2. Comparison of Various Computational Techniques

Some of the more commonly used methods for solving optimal control problems are dynamic programming and the techniques associated with the use of Pontryagin's maximum principle. The latter transforms the problem into a system of differential equations with mixed boundary conditions. One of the most powerful approaches for solving systems of differential equations with mixed boundary conditions is the shooting technique. This method guesses the unknown conditions at one end of the independent variable range and then solves an initial value problem with those conditions. The solutions at the other end of the range are then compared with the known conditions there. If they do not agree, new guesses are made, and the procedure is repeated until the given conditions at both end points are satisfied within an acceptable error. This iteration may be eliminated for systems of linear differential equations. Therefore, the problem can be solved by normal initial value solution techniques, such as the fifth-order Runge–Kutta method and a predictor–corrector method, where the prediction portion is eliminated by making use of the linearity of the problem.

The method of this chapter was compared with the dynamic programming, Runge–Kutta, and corrector methods in several problems. The comparison considered the amount of computer time and space required by each method to obtain a given degree of accuracy. Only one example is given here, and the general results are discussed.

Consider the demonstration problem

$$\dot{X}_1(t) = -X_1(t) + X_2(t) + Z_1(t) - \tfrac{1}{2}Z_2(t)$$

$$\dot{X}_2(t) = 2X_1(t) - \tfrac{1}{2}X_2(t) + \tfrac{1}{4}Z_1 + \tfrac{1}{4}Z_2(t)$$

$$J = \int_0^t [X_1^2(t) + X_2^2(t) + Z_1^2(t) + Z_2^2(t)]\, dt$$

$$X_1(0) = 1.0, \qquad X_1(1) = 0.0$$

and

$$X_2(0) = -1.048, \qquad X_2(1) = 0.0$$

where $X_1(t)$ and $X_2(t)$ are deviations from desired state points, and $Z_1(t)$ and $Z_2(t)$ are the control variables. Pontryagin's maximum principle is applied to this second-order differential system and the performance index J. The resulting problem is a fourth-order coupled differential system with mixed boundary conditions. The results of the comparison of the various methods on this problem are given in Table 4.6. The accuracy is the number of correct digits and the solution, central processor unit (CPU) time is the required time in seconds on a CDC 6400, and space is the necessary array space.

Table 4.6. Comparison of Various Computational Methods

Accuracy	CPU time	Space	Accuracy	CPU time	Space
	STM			Dynamic programming	
7	0.660	929	0	10.750	616
9	0.733	1112	1	23.778	1131
10	0.830	1313	1	194.940	2161
			1	378.898	3771
	Runge-Kutta			Corrector	
6	3.457	349	2	5.823	1484
6	4.908	349	4	11.677	1484
7	8.498	349	6	40.232	1484
9	13.818	349			1484
10	55.515	349			

For this case, it is obvious that the STM uses considerably less computer time than any of the other methods. Also note that as higher accuracy is required the advantages of STM increase. The reason is that the computational effort grows linearly with required accuracy, whereas the increase in computational time needed by the other methods to increase accuracy is considerable. The Runge-Kutta method required less computer space, but the space required by the STM is not limiting. The advantage of the STM over the other methods is borne out by all of the problems examined. The computation time advantage of the STM over the other methods increases as the dimensionality of the problem increases.

4.10.3. Practical Application of STM for Optimal Control of Nuclear Reactors

In reactor operations, it is often desirable to have a particular neutron flux distribution. Possible examples are to improve the burnup of the core, to combat large xenon oscillations, or to obtain uniform power distributions. The problem of finding the optimal control to obtain a particular neutron flux distribution was treated by Stacey (Ref. 4.8).

The one-group kinetics equations with one delay neutron precursor group and temperature feedback can be written as

$$\frac{1}{v}\dot{\phi}(r, t) = [(1 - \beta)\nu\Sigma_f - \Sigma_a + \nabla \cdot D\nabla]\phi(r, t) + \lambda c(r, t)$$

$$-\frac{\partial \Sigma}{\partial T} T(r, t)\phi(r, t) - \Sigma_r(r, t)\phi(r, t), \tag{4.142}$$

$$\dot{c}(r, t) = \beta \nu \Sigma_f \phi(r, t) - \lambda c(r, t) \qquad (4.143)$$

and

$$\dot{T}(r, t) = C_1 \Sigma_f [\phi(r, t) - \phi(r, 0)] - C_2 T(r, t) \qquad (4.144)$$

where typical values are

v = velocity = 1×10^5 cm/sec
ν = average number of neutrons per fission = 2.5
ϕ = neutron flux
β = delayed neutron fraction = 0.0075
Σ_f = macroscopic fission cross section = 0.1 cm^{-1}
Σ_a = macroscopic absorption cross section = 0.249 cm^{-1}
D = diffusion coefficient = 0.5071 cm, taken as a constant for this so that $\nabla \cdot D\nabla \equiv D\nabla^2$
λ = precursor decay constant = 0.075 sec^{-1}
c = precursor concentration
$\partial\Sigma/\partial T$ = temperature coefficient = -0.00054/cm $\cdot$ °C
Σ_r = external control cross section
$C_1 = 1.2405 \times 10^{12}$ cm$^3 \cdot$ °C
$C_2 = 1.09902956$ s^{-1}.

Stacey (Ref. 4.8) removed the nonlinearity of these equations by defining the last two terms of equation (4.142) as the control $z(r, t)$. The performance index measures the deviation of the flux from the desired flux distribution and the amount of control used. It is written as

$$J = \int_0^{t_f} \int_R \{\alpha_1 [\phi(r, t) - \Phi(r, t)]^2 + \alpha_2 z^2(r, t)\} \, dr \, dt \qquad (4.145)$$

where $\Phi(r, t)$ is the desired neutron flux distribution and α_1 and α_2 are weighting constants, which for this problem are 1.0 and 0.01, respectively. For problems of a larger time scale than the neutron flux time scale, the prompt jump approximation, which sets $\partial\phi/\partial t \simeq 0$, is useful. Therefore, two problems are examined. They are

1. Prompt jump approximation without temperature feedback
2. Prompt jump approximation with temperature feedback

The time t_f for the two problems considered here is 50 sec. Both problems are defined such that the system is initially critical with control.

First the problems are simplified by assuming the reactor is an infinite slab of width $L = 200$ cm in the x-direction. Then the spatial dependence is eliminated by using the expansion functions

$$S_k = \left(\frac{2}{L}\right)^{1/2} \sin\left(\frac{k\pi}{L}\right) x, \qquad k = 1, \ldots, M \qquad (4.146)$$

in equations (4.142) through (4.145). Each state variable is now the sum of the expansion functions weighted by the time functions. An example is

$$\phi(x, t) = \sum_{k=1}^{M} \phi_k(t)S_k(x)$$

The time functions are determined from the equations resulting from inserting the spatial dependence of equation (4.146) into equations (4.142) through (4.145) and canceling the spatial dependence. The resultant equations are

$$\frac{1}{\nu} \dot{\phi}_k(t) = \left[(1 - \beta)\nu\Sigma_f - \Sigma_a - D\left(\frac{k\pi}{L}\right)^2\right]\phi_k(t) + \lambda c_k(t) + z_k(t)$$

$$\dot{c}_k(t) = \beta\nu\Sigma_f\phi_k(t) - \lambda c_k(t)$$

$$\dot{T}_k(t) = C_1\Sigma_f[\phi_k(t) - \Phi_k(0)] - C_2 T_k(t)$$

and

$$J_k = \int_0^{t_f} \{\alpha_1[\phi_k(t) - \Phi_k(t)]^2 + \alpha_2 z_k^2(t)\} \, dt$$

for $k = 1, \ldots, M$. In this system of equations, the amplitude functions ϕ_k, c_k, T_k, and z_K are the unknowns to be solved.

For both problems, we assume that the spatial expansion has two modes (i.e., $M = 2$). Initially, the reactor is critical with a control such that the flux distribution's fundamental mode has an amplitude that is twice that of the first harmonic mode. The desired trajectory $\Phi(x, t)$ is that the amplitude of the fundamental mode maintains its initial value while the amplitude of the first harmonic mode decreases linearly from its initial value to zero at $t = t_f$.

The control cross section $\Sigma_r(x, t)$, which is contained in $z(x, t)$, is found here as a continuous function of x, but it could be transformed into discrete line sources. If the reactor control is simulated by L discrete line sources located at x_l, $l = 1, \ldots, L$, then

$$\Sigma_r(x, t) = \sum_{l=1}^{L} R_l(t)\delta(x - x_l) \tag{4.147}$$

The time function of the control

$$z(x, t) = \sum_{k=1}^{M} z_k(t)S_k(x), \qquad k = 1, \ldots, M$$

is defined by Stacey (Ref. 4.8) as

$$z_k(t) = \int_R \psi_k(x)\left[-\frac{\partial\Sigma}{\partial T} T(x, t)\phi(x, t) - \Sigma_r(x, t)\phi(x, t)\right] dr,$$

$$k = 1, \ldots, M \tag{4.148}$$

with expression (4.147) substituted in. If $L = M$, the $R_1(t)$ may be uniquely solved for. If $L < M$, the first $L - 1$ equation (4.148) and the sum of the last $M - L + 1$ equation (4.148) are solved for $R_i(t)$. If $L > M$, $L - M$ of the functions $R_i(t)$ may be specified arbitrarily. For the continuous function representation,

$$\Sigma_r(x, t) = \frac{-(\partial\Sigma/\partial T)T(x, t)\phi(x, t) + z(x, t)}{\phi(x, t)}$$

yields an expression for the control cross section.

Through the use of the maximum principle, a system of differential equations may be found for each problem. For the prompt jump approximation without temperature, the equations with the definition

$$A_k = \left[(1 - \beta)\nu\Sigma_f - \Sigma_a - D\left(\frac{k\pi}{L}\right)^2\right], \qquad k = 1, 2$$

are

$$\dot{c}_k = (B + EF)c_k + EG\psi_k + ER\Phi_k$$
$$\dot{\psi}_k = (W + SF)c_k + (SG - B)\psi_k + (T + SR)\phi_k$$
$$\qquad\qquad\qquad\qquad\qquad\qquad\qquad\qquad k = 1, 2$$
$$z_k = Fc_k + G\phi_k + R\Phi_k$$
$$\phi_k = \frac{1}{A_k}(\lambda c_k + z_k)$$

where ψ_{ik} = adjoint variables defined by the maximum principle

$E = -\beta\nu\Sigma_f/A_k$

$B = \lambda(E - 1)$

$F = -\alpha_1\lambda/(\alpha_2 A_k^2 + \alpha_1)$

$G = E/(2\alpha_2 + 2\alpha_1/A_k^2)$

$R = -\alpha_1/(\alpha_2 A_k + \alpha_1/A_k)$

$W = 2\alpha_1\lambda^2/A_k^2$

$S = 2\alpha_1\lambda/A_k^2$

$T = 2\alpha_1\lambda/A_k$

and the conditions $c_1(0) = 2.5 \times 10^{12}$, $c_2(0) = 1.25 \times 10^{12}$, $\phi_1(50) = 1.0 \times 10^{14}$, and $\phi_2(50) = 0.0$ are given. The computational time needed to solve this problem by the STM on the CDC 6400 computer was 5.9 sec. The flux distribution is shown in Figure 4.7, and the control cross section is shown in Figure 4.8. The former agrees extremely well with the results reported by Stacey. The latter is a continuous control, whereas Stacey reported two line sources as the control.

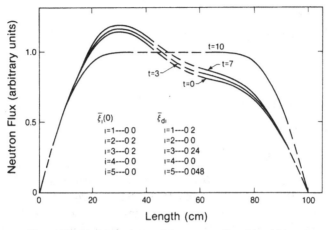

Figure 4.7. Transient shape of the neutron flux $\phi(\omega, t)/\phi_{10}$.

For the prompt jump approximation with temperature feedback, the equations are

$$\dot{c}_k = (B + EF)c_k + EF\Psi_{1,k} + EG_0\psi_{2,k} + ER\Phi_k$$

$$\dot{T}_k = (\lambda + F)E_0c_k - C_2T_k + E_0G\psi_{1,k} + E_0G_0\psi_{2,k} + E_0A_k\phi_k(0) + E_0k\Phi_k$$

$$\dot{\phi}_{1,k} = (W + SF)c_k + (SG - B)\psi_{1,k} + (SG_0 - \lambda E_0)\psi_{2,k} + (X + SR)\Phi_k,$$

$$k = 1, 2$$

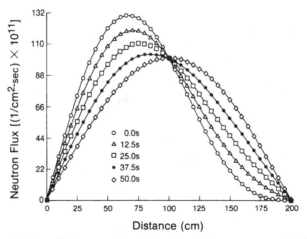

Figure 4.8. Controlled neutron flux distribution without temperature feedback.

$$\dot{\phi}_{2,k} = C_2\psi_{2,k}$$

$$z_k = Fc_k + G\psi_{1,k} + G_0\psi_{2,k} + R\Phi_k$$

$$\phi_k = \frac{1}{A_k}(\lambda c_k + z_k)$$

where $E_0 = -C_1\Sigma_k/A_k$

$$G_0 = E_0/(2\alpha_2 + 2\alpha_1/A_k^2)$$

$$X = 2\alpha_1\lambda A_k$$

and $c_1(0) = 2.5 \times 10^{12}$, $c_2(0) = 1.25 \times 10^{12}$, $T_1(0) = 0.0$, $T_2(0) = 0.0$, $\phi_1(50) = 1 \times 10^{14}$, $\phi_2(50) = 0.0$, $\psi_{2,1}(50) = 0.0$, and $\psi_{2,2}(50) = 0.0$ are given. The computational time needed to solve this problem by the STM on the CDC 6400 computer was 8.2 sec. The flux distribution is shown in Figure 4.10, and the control cross section is shown in Figures 4.9 and 4.11. The flux is almost identical to that of the preceding problem, and the control reveals the effects of the feedback.

4.10.4. Concluding Remarks

The nuclear power industry is becoming a larger and larger contributor to the world's production of electric power. With this growing importance is a need for greater sophistication in the design and operation of nuclear reactors. Therefore efficient methods for solving large-scale optimal control problems are necessary. The main object here was to develop such a method.

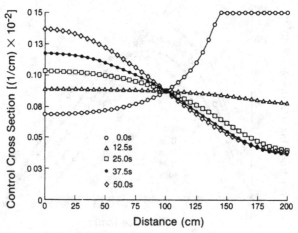

Figure 4.9. Control cross section without temperature feedback.

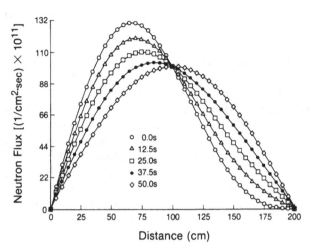

Figure 4.10. Controlled neutron flux distribution with temperature feedback.

In Section 4.10.1, the STM was developed in a systematic manner suitable for computer usage. The method solves the system of coupled differential equations with mixed boundary conditions, which results from the use of the Pontryagin maximum principle on the state equations and the associated performance index of the control problem. In Section 4.10.2 it was shown that, in comparison with other commonly used solution techniques, the STM required considerably less computer time to obtain high-accuracy solutions. It also required less computer space than some of the solution methods. However, the space requirements for the method are

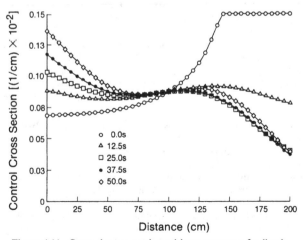

Figure 4.11. Control cross section with temperature feedback.

not large enough to cause any difficulties. The impressively short computer time requirements of the method enable it to be used for large-scale problems.

The STM also has some inherent properties that make it attractive. These were enumerated at the end of Section 4.10.1.

In Section 4.10.3 an optimal control problem relevant to the nuclear reactor field was described. The STM was used to solve this problem in a satisfactory manner with very little computational time.

In conclusion, it can be said that the STM has significant advantages for solving large-scale optimal control problems quickly and efficiently. It may also be used for parametric studies in an extremely efficient manner. Due to some of its inherent properties, parameters may be changed without recalculating the entire solution each time. The method should greatly benefit the design and operation of nuclear reactors.

References

4.1. CHAUDHURI, S. P., "Distributed Optimal Control in a Nuclear Reactor," *Int. J. Control* **16**(5), 927–940 (1972).

4.2. HSU, S., and BAILEY, R. E., "Optimal Control of Spatially Dependent Nuclear Reactors," *Trans. Am. Nucl. Soc.* **10**, 253–260 (1967).

4.3. KAPLAN, S., "The Property of Finality and the Analysis of Problems in Reactory Space-Time Kinetics by Various Modal Expansions," *Nucl. Sci. Eng.* **9**, 357–365 (1965).

4.4. KLIGER, I., "Optimal Control of Space-Dependent Nuclear Reactor," *Trans. Am. Nucl. Soc.* **8**, 233–240 (1965).

4.5. LAZAREVIC, B., OBRADOVIC, D., and CUK, N., "Modal Approach to the Optimal Control System Synthesis of a Nuclear Reactor," *Int. J. Control* **16**(5), 817–830 (1972).

4.6. NIEVA, R., and CHRISTENSEN, G. S., "Optimal Control of Distributed Nuclear Reactors Using Functional Analysis," *J. Optimization Theory Appl.* **34**(3), 445–458 (1981).

4.7. STACEY, W. M., "Optimal Control of Xenon-Power Spatial Transients," *Nucl. Sci Eng.* **33**, 162–170 (1968).

4.8. STACEY, W. M., "Application of Variational Synthesis to the Optimal Control of Spatially Dependent Reactory Models," *Nucl. Sci. Eng.* **39**, 226–235 (1970).

4.9. WANG, P. K. C., "Control of Distributed Parameter Systems," in *Advances in Control Systems*, Vol. 1, edited by C. T. Leondes, pp. 75–110, Academic Press, New York, 1964.

4.10. WIBERG, D. M., "Optimal Feedback Control of Spatial Xenon Oscillations," *Trans. Am. Nucl. Soc.* **7**, 219–225 (1966).

4.11. WIBERG, D. M., "Optimal Feedback Control of Spatial Xenon Oscillations in Nuclear Reactors," *Nucl. Sci. Eng.* **27**, 600–615 (1967).

4.12. WIBERG, D. M., "Optimal Control of Nuclear Reactory Systems," *Adv. Control Syst. Theory Appl.* **5**, 301–388 (1967).

4.13. ASATANI, K., SHIOTANI, M., and HATTORI, Y., "Suboptimal Control of Nuclear Reactors with Distributed Parameters Using Singular Perturbation Theory," *Nucl. Sci. Eng.* **62**, 9–19 (1977).

4.14. ATHAN, M., and FALB, P., *Optimal Control*, McGraw-Hill, New York, 1966.

4.15. NASER, J. A., and CHAMBER, P. L., "An Efficient Solution Method for Optimal Control of Nuclear Systems," *Nucl. Sci. Eng.* **79**, 99–109 (1981).

4.16. LANCZOS, C., Applied Analysis, Prentice-Hall, Englewood Cliffs, New Jersey, 1956.

4.17. STAKGOLD, I., Boundary Value Problems with Mathematical Physics, Vol. 1, MacMillan, London, 1967.

4.18. KELLER, H. B., *Numerical Methods for Two-Point Boundary Value Problems*, Blaisdell, Waltham, Massachusetts, 1968.

4.19. BELLMAN, R. E., and DREYFUS, S. E., *Applied Dynamic Programming*, Princeton University Press, Princeton, New Jersey, 1962.

4.20. CONTE, S. D., *Elementary Numerical Analysis*, McGraw-Hill, New York, 1965.

4.21. KYONG, S. H., "An Optimal Control of a Distributed-Parameter Reactor," *Nucl. Sci. Eng.* **32**, 146–157 (1968).

5

Control of Distributed Reactors in Load-Following

5.1. Introduction (Refs. 5.1–5.28)

If the percentage of nuclear capacity in an electric power system exceeds a certain level, the system must be able to follow a changing level of power load demand within the nuclear power generation portion itself. The effectiveness of a load-following operation in a nuclear reactor depends on the magnitude of the power-level change required and on the duration within which the changes take place.

In a typical daily load cycle in an electric power system, the load demand decreases from 100% to between 50 and 70% of the rated load during 10 P.M. and 1 A.M., holds at the reduced level until 5 A.M. to 6 A.M., and then returns to full power. The total time elapsed during power ramp-up and ramp-down operations that are conducted to follow changing power demand will be between 30 min and 3 h.

Another type of load-following activity is power reduction for a weekend period. Late on Friday night, the core power output is reduced to 50 to 60% of the rated level and remains at the reduced level until Monday morning, when the load begins to rise back to the rated level. The elapsed time at the lower power level is about 55 h and is long enough for xenon to reach an equilibrium state. A severe problem in this case is that the xenon inventory sometimes becomes very low after the upward power ramp, and it consequently becomes very difficult to operate the reactor within the load-line limit.

We summarize the literature on calculational methods for reactor control in the load-following operation.

Love (Ref. 5.17) studied a one-dimensional load-following control problem via a linear programming method. The objective of the control process was to minimize the difference between generated and desired power. Constraints on the axial power distribution offset and controller speed were specified. A nonlinear and a linearized nodal core model with feedback were formulated. The procedure starts from an assumed rod strategy, which is simulated with the nonlinear model to provide the inputs to the linear model through which a new rod strategy is calculated by linear programming. The procedure is then repeated using the new strategy. By iteration of the procedure until the constraints are satisfied, a feasible rod strategy is obtained.

Ebert *et al.* (Ref. 5.18) applied Pontryagin's maximum principle to load-following. Their reactor model was based on a modified one-group theory with Doppler and moderator feedback in one-dimensional geometry. The system and Lagrange equations are not linearized in this procedure; the equations are solved by an iteration method. Assuming a control strategy, the system equations are solved forward in time from the known initial conditions, and the Lagrange equations are solved backward in time from the known final conditions. The control strategy is then modified to minimize the augmented functional at each time point. The process is repeated until no further improvements are achieved.

Karppinen *et al.* (Ref. 5.2) used a multistage mathematical programming (MMP) method to solve a three-dimensional load-following problem. The objective function was expressed as a deviation of the normalized power, xenon and iodine distributions, combined with a control speedy penalty. A nodal core model based on $1\frac{1}{2}$-group diffusion theory was used for simulation, and the control model was derived by linearizing the simulator model and condensing it to a mesh of large zones. The total power, control input, and axial offset were formulated as constraints. The dynamic problem was transformed into a constrained multistage optimization problem, which was solved by a quadratic programming algorithm.

In order to have more spatial dimensions and control freedom, hierarchical control techniques have been used by Beraha and Karppinen (Ref. 5.19) to solve a large-scale problem. The power distribution and xenon-iodine dynamics are decomposed, then costate prediction and mixed methods are used in subproblem coordination. Although extra iterations are needed to achieve a globally optimal solution, the large-scale problem at least becomes solvable. The hierarchical decomposition with respect to time steps is similar to that employed here, but the constraint relations are different.

Lin *et al.* (Ref. 5.4) used a feasible direction method to solve a boiling-water reactor's (BWR) load-following. The objective was to minimize the power mismatch. A one-dimensional model was used with core flow as a control variable. Fuel envelope and load-line constraints were included. Stretching the preconditioned envelope was recommended to increase the flexibility of operation.

Cho and Grossman (Ref. 5.3) formulated their load-following control problem as a linear-quadratic tracking problem, and the resultant two-point boundary value problem (TPBVP) was solved directly by initial value methods. The system equations consisted of the one-dimensional one-group diffusion equation with temperature and xenon feedback, the Xe-I dynamics equations, an energy balance relation for the core, and a state equation of the coolant. The equations were linearized around the steady-state operating point and converted to a lumped parameter system. After application of Pontryagin's maximum principle, the equations were solved by a TPBVP solver.

5.2. Multistage Mathematical Programming (Ref. 5.2)

In this section, the problem of controlling the total power and power distribution in a large reactor core to follow a known time-varying load schedule has been formulated as a multistage optimization problem. This technique is applied to a large pressurized-water reactor (PWR). The control problem is solved subject to hard constraints, which can be applied on total power, control variables and their rate of change, local power densities and their rate of change, and on more global power distribution measures, such as axial and quadrant offsets. Based on a three-dimensional linearized model with some slightly nonlinear features, the optimal control problem is solved by quadratic programming.

5.2.1. Problem Formulation

The control problem is formulated as an objective function and a constraint set. The objective function is based essentially on specified demand power, and xenon and iodine distribution, and these are input data to the problem formulation and assumed to be obtained from design consideration. The total power control is realized by a constraint requiring it to follow the expected load demand. Furthermore, the possible and admissible ways of using the controllers are specified through a set of constraints.

The problem is thus one of finding the control sequence that minimizes the objective function while satisfying all the constraints. The state equations, which link the core state variables to the control variables, are derived by linearizing a nodal $1\frac{1}{2}$-group diffusion theory core model with xenon-iodine dynamics. The nodal equations are condensed to correspond to a mesh of large zones, and only the average power, xenon, and iodine concentrations in these zones are considered in the optimization. The model formulated in terms of these zones is called the *control model*, and its variables are factorized into one factor describing the bulk-level changes and another factor describing the normalized distribution changes. Some nonlinearity is thereby also introduced. The unknown control inputs of the problem will be the control (control rod banks and boron concentration) speeds.

5.2.2. Objective Function

The *objective function* is an integral over the control period (t_0, t_d) and the reactor core volume V:

$$
J = \int_{t_0}^{t_f} dt \int_V dr \bigg(\{ W_p(r, t)[P^n(r, t) - P_d^n(r, t)]^2
$$

$$
+ W_x(r, t)[X^n(r, t) - X_d^n(r, t)]^2 + W_I(r, t)[I^n(r, t) - I_d^n(r, t)]^2 \}
$$

$$
+ \sum_{r=1}^{R} \{ W_{u,r}(t)[U_r(t) - U_{r,d}(t)]^2 + W_{v,r}(t)v_r^2(t) \} \bigg) \qquad (5.1)
$$

where P^n, X^n, I^n = normalized (to core average) power density, xenon concentration, and iodine concentration distributions, respectively

U_r, v_r = values of the control inputs (U_r, $r = 1, \ldots, R$) and their time rates of change (v_r, $r = 1, \ldots, R$)

$W(r, t)$ = weighting functions to be freely chosen

The subscript d is used to indicate the desired value of the variables. Since the normalized distributions are used, they are independent of the bulk levels in variable-load operation, and the objective function thus reflects only the spatial distribution control objective. The xenon and iodine distributions are included because they are the prime causes of power oscillations. The time- and position-dependent weighting factors can be used to express the relative importance of different terms during different parts of a load cycle.

The control values and speeds in the objective function can be used to specify desired control rod positions and boron concentration values, by means of which the total control effort can be minimized and soft priorities can be assigned for different control variables. Hard constraint must be treated through the constraint set.

The objective function is extremized subject to satisfying the constraint set given by

$$
\int_V dr\, P(r, t) = P_{\text{tot},d}(t) \qquad (5.2)
$$

$$
P(r, t) \le P_{\max}(r, t), \qquad r \in V \qquad (5.3)
$$

$$
\dot{P}(r, t) \le \dot{P}_{\max}(r, t), \qquad r \in V \qquad (5.4)
$$

$$q^l_{min}(t) \leq \int_V dr\, a^l(r,t) P(r,t) \leq q^l_{max}(t), \qquad l = 1,\ldots,L \qquad (5.5)$$

$$U_{r,min}(t) \leq U_r(T) \leq U_{r,max}(t), \qquad r = 1,\ldots,R \qquad (5.6)$$

$$v_{r,min}(t) \leq v_r(t) \leq v_{r,max}(t), \qquad r = 1,\ldots,R \qquad (5.7)$$

Control of the total power according to the demand $[P_{tot,d}(t)]$ is realized by equation (5.2) and limitations on the local power density, $P(r,t)$, and its rate of change in inequalities (5.3) and (5.4). More global constraints on the power density distribution can be specified with inequality (5.5). A constraint on the axial offset is, for example, obtained with

$$a(r,t) = \begin{cases} -\dfrac{100}{P_{tot}(t)}, & r \in V_u \\[2ex] +\dfrac{100}{P_{tot}(t)}, & r \in V_l \end{cases} \qquad (5.8)$$

where V_u and V_l are the upper and lower core halves, respectively. Constraints on quadrant peakings can be specified similarly. Many operational limitations, technique specifications, and criteria on mechanisms leading to fuel defects can be correlated with $P(r,t)$ and $\dot{P}(r,t)$ and thus are covered by the above constraints. The control value and speed constraints, inequalities (5.6) and (5.7), can be used to define their operating ranges and speeds.

5.2.3. Reactor Core Model

The control models or the state equations used in the optimization have been derived from the core simulator model, which has been used in the simulation studies. Therefore, the simulator model, which is a nodal diffusion code with xenon-iodine dynamics, is first described.

5.2.3.1. Simulator Model

In the $1\frac{1}{2}$-energy-group formalism, neutron diffusion can be represented by

$$-M^2 \nabla^2 \phi_1(r) + \phi_1(r) = \frac{1}{K} K_\infty \phi_1(r) \qquad (5.9)$$

where ϕ_1 = fast neutron flux
$\quad M$ = migration length
$\quad k_\infty$ = local infinite multiplication factor
$\quad k$ = eigenvalue

Since the thermal flux has been eliminated through the asymptotic thermal-to-fast-flux ratio approximation, the power density can be assumed to be proportional to the fast flux with a position-dependent factor. Equation (5.9) can be discretized by dividing the core into N_n nodes and by considering the average values of the variables and parameters in these nodes. It can be further transformed into an integral equation and written as

$$LP = \frac{1}{k} NP \tag{5.10}$$

where P is a vector of nodal power densities and L and N are the neutron destruction and production operators, respectively, defined as

$$L_{nm} = \left(\frac{1}{k_n} + \sum_{l \in V_R} \omega_{nl}\right) \delta_{nm} \tag{5.11}$$

$$N_{nm} = \left(1 - \sum_l \omega_{nl}\right) \delta_{nm} + \omega_{mn} \tag{5.12}$$

with

$$\omega_{mn} = \frac{M_{mn}}{2h_{mn}} \left[\frac{k_m - k_n}{2k_m}\right] + \frac{M_{mn}^2}{h_{mn}^2} \left[\frac{k_m + k_n}{2k_m}\right] \tag{5.13}$$

$$M_{mn} = \frac{m_m + M_n}{2} \tag{5.14}$$

and where M_n and k_n are the nodal values of M and k_∞, h_{mn} is the distance between center points of nodes m and n, δ_{mn} is the Kronecker delta, and $l \in V_R$ refers to the nodes in the reflector. The spatial coupling coefficients ω_{mn} describe the neutron diffusion, and this above form is derived by assuming continuity of the absorption rate. The second term in equation (5.11) describes the leakage from core to reflector. The diffusion of thermal neutrons back into the core is accounted for by a later correction.

Parameters k_∞ and M, and therefore operators L and N, depend on such variables as the xenon distribution, fuel temperature, moderator density, boron concentration, and control rod configuration in a nonlinear way. The simulator has a full description of the thermal feedback. The xenon-iodine dynamics are described by the nodal equations

$$\frac{dX_n}{dt} = -\lambda_x X_n + \lambda_I I_n - \Gamma_n P_n X_n + \gamma_x^* P_n \tag{5.15}$$

$$\frac{dI_n}{dt} = -\lambda_n I_n + \gamma_I^* P_n, \qquad n = 1, \ldots, N_n. \tag{5.16}$$

where the xenon and iodine yields from fission, γ_x^* and γ_I^*, as well as the xenon microscopic absorption cross section, Γ_n, have been modified to allow the use of power density in the equations. Here, Γ_n depends on moderator density, boron concentration, and presence of control rods in the node.

The xenon transient calculation with the simulator employs an iterative solution of the static diffusion equation simultaneously with iterative time integration of the xenon-iodine dynamics equations with time-varying power. That corresponds to a fully implicit time integration scheme, making possible large time steps and a realistic description of control rods moving with a constant speed. The description of the rod and boron control system is quite versatile, allowing different control rod grouping, criticality search with respect to different parameters, specification of priorities, and admissible operation ranges and speeds for controllers, etc.

5.2.3.2. Control Model Spatial Condensation

Simplifications are introduced into the core model to reduce the optimization problem to a computationally manageable size. The simplifications are with respect to system order and nonlinearities. The first aspect is treated in this section. The simulator model has $2N_n$ dynamic state variables, equations (5.15) and (5.16), and N_n nonlinear algebraic constraints, equation (5.10). In a realistic three-dimensional case, the system order would easily be on the order of thousands and thus far beyond the computational capability of most optimization methods. The approach suggested here is to collapse or condense the nodal equations to a set of "zonal" equations of a considerably reduced order, geometrically corresponding to a division of the core into N_z large zones, each consisting of several nodes.

Consider first the power equation, equation (5.10). For each node n we have

$$\frac{1}{k_n} P_n = \frac{1}{k} \sum_{m\Sigma A_n} \omega_{mn} P_m, \qquad n = 1, \ldots, N_n \qquad (5.17)$$

where the nodal k_n's are corrected for eventual boundary effects and $m\Sigma A_n$ refers to the six nodes m that are adjacent to node n and the node n itself. The average power density in zone i, with volume V_i, of the N_z above introduced larger zones is thus

$$P_i = \frac{1}{V_i} \sum_{n \in V_i} V_n P_n, \qquad i = 1, \ldots, N_z \qquad (5.18)$$

Collapsing all equations of equation (5.17), weighted by volumes V_n, corresponding to nodes n that belong to zone i ($n \in V_i$) into one equation yields

$$\frac{1}{k_i} P_i = \frac{1}{k} \sum_{j \in A_i} \omega_{ji} P_i, \qquad i = 1, \ldots, N_z \qquad (5.19)$$

with

$$\frac{1}{k_i} = \frac{1}{P_i} \sum_{n \in V_i} V_n \frac{1}{k_n} P_n \tag{5.20}$$

$$\omega_{ji} = \frac{1}{P_j} \sum_{n \in V_i} V_n \sum_{m \in A_{n,j}} \omega_{mn} P_m \tag{5.21}$$

where $m \in A_{n,j}$ refers to those nodes m that are adjacent to node n and within zone j. Equation (5.21) also applies to the self-coupling coefficient ω_{ii}, with the set $A_{n,j}$ defined accordingly. When the same condensation is done for the xenon and iodine equations, we obtain a set of equations for the zonal variables that are formally identical with equations (5.15) and (5.16). The coefficients will be weighted averages, e.g.,

$$\Gamma_i = \frac{1}{V_i P_i X_i} \sum_{n \in V_i} V_n \Gamma_n X_n P_n \tag{5.22}$$

Once the converged nodal solution is found, the above condensation technique can be used to obtain the zonal equations. The resulting zonal model corresponds to the particular core state at which the condensation is done, and the zonal model is consistent with the nodal model at that state. The zonal model parameters are based on the distributions prevailing within the zones at that state. If the zonal model is then used to simulate the core behavior, results can be expected to deviate from those of the nodal model by an amount that increases with increasing difference in the intrazonal distributions.

5.2.3.3. Control Model Linearization

The nonlinearities of the present system are

1. The power-xenon product term
2. The dependence of the group constants on power density through thermal feedbacks
3. The neutron flux–control absorber concentration product term

These three nonlinearities will essentially be linearized.

For a critical core ($k = 1$), the N_z zonal equations, equations (5.19), can be written $LP = NP$, which can be differentiated with respect to time to yield

$$\dot{L}P + L\dot{P} = \dot{N}P + N\dot{P} \tag{5.23}$$

where L and N are time-varying through the dependence on xenon distribution, thermal feedback, and control. Neutronic, thermal, and hydraulic feedbacks are considered to be prompt.

If the spatial coupling coefficients, equation (5.21) are assumed to be constant ($\dot{N} = 0$), and if only the variations in L are taken into account as

$$\dot{L} = \frac{\partial L}{\partial X}\dot{X} + \frac{\partial L}{\partial C}\dot{C} + \left(\frac{\partial L}{\partial P}\right)\dot{P} \tag{5.24}$$

then we can rewrite equation (5.23) as

$$\left[L - N + \frac{\partial L}{\partial P}\right]\dot{P} = -\left(\frac{\partial L}{\partial X}\right)P\dot{X} - \left(\frac{\partial L}{\partial C}\right)P\dot{C} \tag{5.25}$$

Here, the power (P), xenon (X), and control poison (C) distributions, as represented by the zonal averages, are written as diagonal matrices, as are the partial derivatives. These have the diagonal elements

$$\left(\frac{\partial L}{\partial X}\right)_{ii} = -\frac{1}{k_i^2}\frac{\partial k_i}{\partial X_i} \tag{5.26}$$

$$\left(\frac{\partial L}{\partial C}\right)_{ii} = -\frac{1}{k_i^2}\frac{\partial k_i}{\partial C_i} \tag{5.27}$$

$$\left(\frac{\partial L}{\partial P}\right)_{ii} = -\frac{1}{k_i^2}\frac{\partial k_i}{\partial T_i}\frac{\partial T_i}{\partial P_i} \tag{5.28}$$

In equation (5.28), T_i represents the thermal feedback due to both fuel and moderator temperatures. Using the notation

$$A_s = \left[L - N + \frac{\partial L}{\partial P}\right] \tag{5.29}$$

we can state that an explicit relation between the power distribution changes and the xenon and control poison distribution changes can be obtained if A_s^{-1} exists.

For a critical core, the matrix $L - N$ has eigenvalues in the interval $(-\infty, 0)$, and with the largest eigenvalue equal to zero the matrix is singular. However, the A_s matrix is modified by $\partial L/\partial P$, which has elements on the order of magnitude of the Doppler feedback coefficient. The eigenvalues are thereby shifted to more negative values, leading to a negative definite A_s matrix with an existing inverse. Since the size of the A_s matrix is reduced to $N_z \times N_z$ by the condensation procedure, it is also invertible in practice.

The control poison distribution changes due to the changes of R control variables $U_1, \ldots, U_R$ are described by a $N_Z \times R$ transformation matrix D,

$$\dot{C} = D\dot{U} \tag{5.30}$$

which is a discontinuous function of U.

The core state is described by the state vector

$$Y^T = [P^T X^T I^T U^T] \tag{5.31}$$

which has dimension $3N_z + R$. The zonal model equations, equations (5.25), and those corresponding to equations (5.15) and (5.16) are now linearized around a reference state Y_0 as follows:

$$Y(t) = y(t) + Y_0 \tag{5.32}$$

yielding

$$p(t) = C_{px}x(T) + C_{pu}u(T) \tag{5.33}$$

$$\begin{bmatrix} \dot{x}(t) \\ \dot{i}(t) \end{bmatrix} = \begin{bmatrix} A_{xx} & A_{xi} \\ A_{ix} & A_{ii} \end{bmatrix} \begin{bmatrix} x(t) \\ i(t) \end{bmatrix} + \begin{bmatrix} C_{xu} \\ C_{iu} \end{bmatrix} u(t) + \begin{bmatrix} \dot{X}_0 \\ \dot{I}_0 \end{bmatrix} \tag{5.34}$$

with

$$C_{px} = -A_s^{-1}\left(\frac{\partial L}{\partial x}\right) P_0 \tag{5.35}$$

$$C_{pu} = -A_s^{-1}\left(\frac{\partial L}{\partial C}\right) P_0 D \tag{5.36}$$

$$A_{xx} = -\Lambda_x - \Gamma P_0 + (\gamma_x + \Gamma X_0) C_{px} \tag{5.37}$$

$$A_{xi} = \Lambda_I \tag{5.38}$$

$$A_{ix} = \gamma_I C_{px} \tag{5.39}$$

$$A_{ii} = -\Lambda_I \tag{5.40}$$

$$C_{xu} = (\gamma_x + \Gamma X_0) C_{pu} \tag{5.41}$$

$$C_{iu} = \gamma_I C_{pu} \tag{5.42}$$

Equation (5.34) describes the system dynamics, and was derived by eliminating the power density by using equation (5.33). Equation (5.33) thus represents only an algebraic feedback relation through which the power distribution changes are obtained when $x(t)$ and $u(t)$ are known. The linearized model parameters (5.35) through (5.42) depend on the linearization state power and xenon distributions P_0 and X_0 as well as on the partial derivatives of equation (5.24) directly and through the matrix A_s^{-1}, which again depends on the inverse of the spatial coupling coefficient matrix. The "core response matrices" C_{px} and C_{pu} link the effects of poison (xenon and control absorbers) distribution changes directly to changes in xenon and iodine distributions by eliminating the neutron flux distribution through which the physical process works.

5.2.3.4. Control Model Factorization

The state vector $Y(t)$ is further factorized into two parts, one describing the amplitude and the other the distribution:

$$Y(t) = Y_L(t) y_N(T) \tag{5.43}$$

where $Y_L(t)$ is a diagonal matrix with N_z first diagonal elements equal to the core average power density and the following elements equal to the core average xenon and iodine concentrations. The corresponding amplitude factor for control is one. The vector $y_N(t)$ contains the normalized power, xenon, and iodine distributions and the actual control values.

The deviation from the reference state is now

$$y(t) = Y_L(t)y_N(t) - Y_{L,0}y_{N,0}$$
$$= Y_L(t)\Delta y_N(t) + \Delta Y_L(t)y_{N,0} \tag{5.44}$$

with

$$\Delta y_N(t) = y_N(t) - y_{N,0} \tag{5.45}$$

$$\Delta Y_L(t) = Y_L(t) - Y_{L,0} \tag{5.46}$$

Here, $\Delta y_N(t)$ describes the change in the normalized distributions, and $\Delta Y(t)$ the changes in the amplitudes. Substitution of the xenon and iodine parts of equation (5.44) into equation (5.34) gives

$$\dot{z}(t) = Az(t) + Cu(t) + f(t) \tag{5.47}$$

with

$$z(t) = Y_L(t)\Delta y_N(t) \tag{5.48}$$

$$f(t) = [A\Delta Y_L(t) - \Delta \dot{Y}_L(t)]y_{N,0} + d \tag{5.49}$$

where A, C, and d are the corresponding elements of equation (5.34). A new computational state variable is defined by equation (5.48). System matrices A and C are time-invariant, and the vector f depends on the amplitude factors $Y_L(t)$. The amplitude factors can be calculated by a nonlinear point model obtained by further condensing the zonal xenon-iodine equations:

$$\dot{X}_L(T) = -\lambda_x X_L(t) - \langle \Gamma y_N \rangle X_L(T)P_L(T) + \langle \gamma_x^* P_N \rangle P_L(t) \tag{5.50}$$

$$\dot{I}_L(t) = -\lambda_I I_L(t) + \langle \gamma_I^* P_N \rangle P_L(t) \tag{5.51}$$

where the $\langle \ \rangle$ terms are distribution-weighted model parameters and where the distribution time derivatives are neglected. By using the reference point distributions $y_{N,0}$ for weighting and the total power demand $P_{tot}(t)$ of equation (5.2) as $P_L(t)$ for $t \in (t_0, t_f)$, we can easily integrate equations (5.50) and (5.51).

The control model of equation (5.47) is thus linear with respect to the normalized distribution changes, but has the nonlinear feature introduced by the use of the amplitude factors and the point model. It is also possible to add other known effects that influence the power distribution control into the $f(t)$-vector. One such effect is the variation of the coolant inlet temperature $T_{in}(t)$. Considering the fuel and moderator temperature feedbacks separately and together leads to a nondiagonal $(\partial L/\partial P)$ matrix in equation (5.24) and will be reflected through the A_s^{-1} matrix. The f-vector will then be appended with a term describing the deviation of $T_{in}(t)$ from a reference value.

5.2.3.5. Description of Control Variables

The linearization of the bilinear control term $P(r, t) \cdot C(r, t)$, where $C(r, t)$ is the control poison distribution, is somewhat problematic since both $P(r, t)$ and $C(r, t)$ are space- and time-dependent and, in particular, because $C(r, t)$ is the unknown to be determined.

In equation (5.33) the effects of control on the power distribution were described by the core response matrix

$$C_{pu} = -A_s^{-1}\left(\frac{\partial L}{\partial P}\right) PD$$

Going back to the nodal core representation, the element $c_{i,r}$ of C_{pu}, connecting the movement of controller r to the power change in zone i, would be

$$c_{i,r} = \sum_{n \in V_i} V_n \sum_{m=0}^{N_n} [(A_s^{-1})_{nm} \xi_m P_m] D_{m,r} \qquad (5.52)$$

where $\xi_m = (\partial L/\partial C)_{mm}$ of equation (5.27). The controller geometry matrix D, defined in equation (5.30), has nonzero elements corresponding to those nodes m only where the effect of controller r is changing—e.g., where a control rod tip is moving:

$$D_{m,r} = d_{m,r}\delta(m, r) \qquad (5.53)$$

where $\delta(m, r)$ is unity for the above identified nodes n and zero elsewhere, and where d_{mr} converts the controller movement into control poison change.

The nodal A_s^{-1} matrix in equation (5.52) acts as a discretized Green's function by reflecting the effect of the local reactivity change in node m on the power density in all the other nodes. In terms of the zonal variables, equation (5.52) is approximated by

$$(C_{pu})_{i,r} = \sum_{j=1}^{N_i} [(A_s^{-1})_{ij} \xi_j P_j] d_{jr} \delta(j, r) \qquad (5.54)$$

where the zonal A_s^{-1} matrix is already defined through the zonal coupling coefficients, equations (5.21) and (5.29). The zonal ξ_j's will thus be

$$\xi_j = \frac{1}{V_j P_j} \sum_{n \in V_i} V_n P_n \xi_n \qquad (5.55)$$

The linearization of the model appears in two respects. First, the A_s matrix is evaluated at the reference state and its changes are thereafter not taken into account, and the zonal ξ's are calculated by using the reference point power distributions. Second, the D matrix must also be evaluated at the reference point control rod configuration, and its nonlinear dependence on the control rod movement must be neglected. These two limitations in the control rod description are common to all kinds of linear representations.

The core response matrix, given by equation (5.36), can be written as $C_{pu} = BPD$, and depends on the power distribution P. Since the total power demand $P_{tot}(t)$ is assumed to be known, one can approximate $P(t)$ by

$$P(t) = \frac{P_{tot}(t)}{P_{tot,0}} P_{N,0}$$

and write

$$C_{pu}(t) = \frac{P_{tot}(t)}{P_{tot,0}} Bp_{N,0}D \tag{5.56}$$

thereby introducing some nonlinearity into the control description.

5.2.4. Numerical Solution

5.2.4.1. Time Discretization

The total control period (t_0, t_f) is divded into K time steps

$$\Delta t_k = t_k - t_{k-1}, \qquad k = 1, \ldots, K$$

which can be of different lengths. The control model equations to be discretized are equations (5.33), (5.47), (5.50), and (5.51). Integration of equation (5.47) over Δt_k yields

$$z(t_k) = \exp(-A_d \Delta t_k) z(t_{k-1})$$
$$+ \int_{t_{k-1}}^{t_k} dt \, \exp[-A_d(t - t_{k-1})][A_{fz}(t) + C^* P_{tot}(t)u(t) + f(t)]$$
$$\tag{5.57}$$

where the A matrix has been split into a diagonal part A_d and an off-diagonal part A_f, and the power-level dependence of C has been indicated explicitly. All the time-dependent variables of the integrand are assumed to change linearly with time. The time integration scheme is thus implicit and corresponds to an exponentially fitted Padé $(1, 1)$ method. The linear time dependence of $u(t)$ means constant control speeds v_k over Δt_k and results in a change $\Delta u(k) = v(k)\Delta t_k$ in the control variable values. When the power equation (5.33) is also modified into the factorized form and all the discretizations are performed, the following equation is obtained:

$$z(k) = \begin{bmatrix} 0 & G_{px}^k & G_{pi}^k & G_{pu}(P_L) \\ 0 & G_{xx}^k & G_{xi}^k & G_{xu}^k(P_L) \\ 0 & G_{ix}^k & G_{ii}^k & G_{iu}^k(P_L) \\ 0 & 0 & 0 & 1 \end{bmatrix} z(k-1)$$

$$+ \begin{bmatrix} C_{p\Delta u}^k(P_L) \\ C_{x\Delta u}^k(P_L) \\ C_{i\Delta u}^k(P_L) \\ I \end{bmatrix} \Delta u(k) + \begin{bmatrix} g_p^k(Y_L) \\ g_x^k(Y_L) \\ g_i^k(Y_L) \\ 0 \end{bmatrix} \quad (5.58)$$

$$Y_L(k) = F[P_L(k), Y_L(k-1), Y_0^k] \quad (5.59)$$

Equation (5.59) represents the points model. Dependencies of the spatial model parameters on the power level P_L and generally on the amplitude factors Y_L are denoted explicitly. The state vector z is defined by equation (5.48), and the core state is obtained as

$$Y(k) = z(k) + \Delta Y_L(k)y_{N,0}^k + Y_0^k \quad (5.60)$$

The objective function variables can be expressed in terms of the computational state variables $z(k)$, as well as the desired distributions $z_d(k)$. The objective function can thus be discretized to

$$J_K = \sum_{k=1}^{K} \{[z(k) - z_d(k)]^T W^k [z(k) - z_d(k)] + \Delta u^T(k) R^k \Delta u(k)\} \quad (5.61)$$

The constraints are replaced by their discrete time equivalents, and finite differences are used for the time derivatives.

5.2.4.2. Transformation to the Quadratic Programming (QP) Form

The dynamic control problem has now been transformed into a constrained multistage optimization problem over K time steps. The decision variables or control inputs to be determined are the $\Delta u(k)$ vectors for $k = 1, \ldots, K$. The optimization problem is further transformed into standard QP form.

Once the total power demands and thus $P_L(k)$, $k = 1, \ldots, K$, is known, the point model yields the amplitude factors $Y_L(k)$, and the control model parameters for all the time steps can be evaluated. A different control model can be used for each time step corresponding to different reference states. By recursive use of the state equation, equation (5.58), any state $z(k)$ can be expressed as a function of the initial state $z(0)$ at t_0 and the control sequence $\Delta u(l)$, $l = 1, \ldots, k$:

$$z(k) = \sum_{l=1}^{k} T(k, l)\Delta u(l) + \left(\prod_{i=1}^{k} G^i \right) z(0) + \sum_{l=1}^{k} \left(\prod_{i=l+1}^{k} G^i \right) g^l \quad (5.62)$$

with

$$T(k, l) = G^k G^{k-1} \cdots G^{l+1} C^l \quad (5.63)$$

Substitution of these expressions into the objective function, equation (5.61), gives

$$J_K = C^T X + \tfrac{1}{2} X^T Q X \quad (5.64)$$

with

$$X^T = [\Delta u^T(l) \cdots \Delta u^T(k) \cdots \Delta u^T(K)] \quad (5.65)$$

$$c_k = \sum_{l=k}^{k} T(l, k) W^l \Delta^l, \qquad k = 1, \ldots, K \quad (5.66)$$

$$\Delta^k = \left(\prod_{i=1}^{k} G^i \right) z(0) + \sum_{l=1}^{k} \left(\prod_{i=l+1}^{k} G^i \right) g^l \quad (5.67)$$

$$Q_{k,m} = \sum_{l=\max\{k,m\}}^{K} [T^T(l, k) W^l T(l, m) + \delta_{k,m} R^l],$$

$$k = 1, \ldots, K, \quad m = 1, \ldots, K \quad (5.68)$$

where $Q_{k,m}$ and C_k are $(R \times R)$- and R-dimensional submatrices and subvectors of Q and c, respectively. Similarly, the constraints can be expressed as a function of the control vector X, and the constraint set can be transformed to

$$A_1 X = b_1 \quad (5.69)$$

$$A_2 X \geq b_2 \quad (5.70)$$

Equation (5.69) represents the total power and possibly other equality constraints, and inequality (5.70) represents constraints, both those on control variables on the reactor state.

5.2.4.3. Solution of the QP Problem

The QP problem posed in equations (5.64), (5.69), and (5.70) is solved by standard QP algorithms. A QP problem has a unique solution if the Q-matrix is positively semidefinite and the feasible region formed by the constraints is not empty. The first requirement is fulfilled by choosing nonnegative weights W. The feasible region may be empty if constraints that are too restrictive are specified on the control and state variables, indicating simply that it is impossible to follow the load demand. A solution could then be found by relaxing some of the constraints. Reformulation of the problem to allow the total power to vary between an upper and lower limit would also be possible with minor modifications.

The computation QP problem size is determined, on one hand, by the dimension of X, $K \cdot R$, which does not depend on N_z, the number of zones, and, on the other hand by the number of constraints. Since some of the constraints can, in practice, be redundant, they can be activated in groups or even individually. In practice, the number of constraints is usually the limiting factor, not $K \cdot R$. In general, linear programming (LP) algorithms are more efficient in constraint handling than are QP algorithms. In an LP formulation of the control problem, the constraint set would be the same but the objective function would have to be linear. The squaring of the deviation in equation (5.1) would be replaced by the absolute value, or a minimax criterion could be applied.

5.2.5. Operational Use

Figure 5.1 shows the operational use of the method as used in simulations and as contemplated to be eventually implemented on a reactor plant.

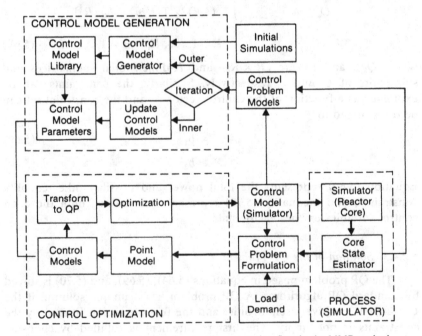

Figure 5.1. Proposed operational use and computation flow in the MMP method.

The main control loop is indicated with a heavy line, and thin lines denote auxiliary loops and information flow.

5.2.5.1. Use in Simulation

Simulation studies with the method start with a guessed initial simulation run with the nonlinear nodal code (the simulator), from which the core state and main model parameters, including the spatial coupling coefficient matrix A_s are stored. The data are used by the control model generator to calculate the discrete time zonal control model parameters for the spatial and point models [parameters of equation (5.58)].

These control model parameters for different reference states are stored to form a control model library. A control study starts by defining the load demand [$P_L(k)$'s in equations (5.58) and (5.59)] for a period of time ahead. The control problem formulation takes into account the load demand and other operating requirements and sets up the control problem by defining the control period and dividing it into time steps, assigning control models to them, by specifying the objection function and constraints and initial condition state vector. The point model [equations (5.50) and (5.51)] then calculates the amplitude factors, whereby the control models are complete, and the control problem transformation into QP form [equations (5.62) through (5.70)] can be performed. The optimization yields the control sequence, and the control is simulated with the control models to get the control trajectory. At that point, there are two possibilities. Either the control can be simulated with nonlinear simulator, or the control sequence can be improved in inner iterations by updating the control models [equation (5.71)] along the control trajectory. The latter case yields new control model parameters and is followed by a new transformation and optimization. After one or more inner iterations, the control is simulated with the simulator.

After the simulation, one can either stop, continue with a new load cycle, or still improve the control in an outer iteration loop. In the outer iteration loop, a relinearization along the simulator trajectory takes place; i.e., new control models are generated using the simulator calculated core states as the reference states. Each outer iteration can include several inner iterations. The simulator thus has a dual role: it is used as a part of the overall control algorithm and for simulating the process to be controlled.

When continuous on-line operation of the control system is simulated, the control for a period of time ahead is always calculated, and then the resulting control is simulated. The decision on the control problem setup, control models, and whether to use the inner and outer iterations would be done in the blocks' control problem formulation and control problem models. If, for example, the expected load demand changes or the core behavior differs considerably from that predicted by the control system, these modules should initiate the calculation of a new control sequence before the previous control is fully executed. The core state estimation includes the estimation of the power, xenon, and iodine distributions, which in the simulations have been assumed directly measurable.

In an actual implementation on a reactor, an on-site simulator would still be useful. After some inner iterations, the control sequence could be tested with the simulator, as indicated in parentheses in Figure 5.1, and, if necessary, outer iterations could be performed. The on-site simulator could be used for core-following calculation in the xenon load-following time scale by utilizing core measurements. That could also solve the core state estimation problem.

The amount of on-line computations required for implementing the above outlined control scheme could vary considerably. The most general case would be to perform all the above-mentioned tasks on-line, including the model generation and various iterations. An implementation scheme requiring least on-line computation would include only the control optimization part, and the operational loop would then be that indicated by the heavy arrows in Figure 5.1. Typical execution time for the research versions of the programs constituting the CONTROL OPTIMIZATION block is about 60 CPU seconds on a CDC Cyber 74 large scientific computer. That calculation would have to be repeated once for each control period, which could be several hours.

5.2.5.2. Iterative Updating of Control Models

In the inner iterations, most of the control model parameters could, in principle, be updated along the control trajectory. The core response matrix C_{pu}, equation (5.56), is updated on iteration cycle i according to

$$C_{pu}^i = rC_{pu}^{i-1} + (r-1)C_{pu}(z^i) \tag{5.71}$$

where $C_{pu}(z^i)$ is the core response matrix at state z^i and r is an underrelaxation factor. The term $C_{pu}(z^i)$ could be approximated as follows:

$$C_{pu}(z^i) = \frac{P_{\text{tot}}^i}{P_{\text{tot},0}}[B(z_0) + \Delta B(z^i)]p_N^i \cdot D(u^i) \tag{5.72}$$

with

$$\Delta B(z^i) = \sum_j \left[A_s^{-1} \frac{\partial}{\partial Z_j}\left(\frac{\partial L}{\partial C}\right) - A_s^{-1}\frac{\partial A_s}{\partial Z_j} \cdot A_s^{-1}\left(\frac{\partial L}{\partial C}\right) \right]_{z=z_0} \Delta Z_j^i \tag{5.73}$$

$$\Delta z^i = z^i - z_0 \tag{5.74}$$

where j runs over all components of z; $D(u^i)$ is the controller geometry matrix of equation (5.53) for the controller configuration u^i. By the above technique of updating C_{pu}, we avoid finding the inverse of the A_s matrix but we require the sensitivity coefficient matrices $\partial A_s/\partial z_j$ at $z = z_0$. In the simulations reported below, only $D(u^i)$ has been updated; i.e., $\Delta B = 0$ and $p_N^i = p_{N,0}$.

5.2.6. Results

The results presented in Section 5.2.6.1 illustrate the control rod description with a one-dimensional example. In Sections 5.2.6.2 and 5.2.6.3, results from three-dimensional control studies are reported. In all three cases, the simulator nodes were identical with the control model zones; i.e., the condensation procedure was not employed. The core model data, dimension, and power density were representative of a 3600-MW (thermal) class PWR core. However, the models were tuned to be fairly strongly unstable with respect to axial and, in the three-dimensional case, axial and azimuthal xenon oscillations.

5.2.6.1. Control Rod Description

The one-dimensional core model had 13 axial zones and only one very strong control rod bank and boron control. Figure 5.2 depicts the change in the normalized axial power distribution resulting from a 50-cm control rod bank movement with constant speed in 1 h. The reactivity effects were not compensated for, and thus the total power was reduced from 100 to 66%. The resulting "exact" normalized distribution is calculated with the simulator. For comparison, both the resulting distribution as calculated by the MMP control model and by a corresponding linearized model (LIN) without the factorization and the power-level-dependent core response matrix are presented. For the MMP model, the inner iteration cycle was used to update the $D(u^i)$ matrix.

The MMP model gives significantly better results than does the LIN model. Both models overestimate the local power depression around the moving rod tip. This is due to the linearization of the bilinear control term. Therefore, the effect of the control poison insertion on the power distribution is calculated as if the system properties and the power distribution would not change due to the rod movement. That can be seen more exactly from equations (5.56) and (5.72). The Green's function matrix B and normalized power distribution p_N^i are taken to be constant, although they change concurrently with rod movement. That is, however, a nonlinear effect, and can be taken into account only iteratively, either in the inner iterations,

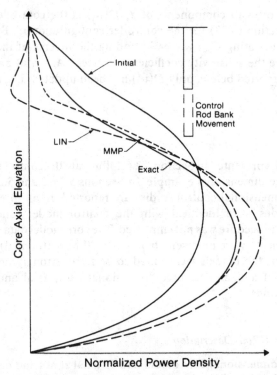

Figure 5.2. Change in the normalized power distributed due to a 50-cm insertion of a strong control rod bank in 1 h (no reactivity compensation) as calculated by the simulator (EXACT), MMP model, and a purely linear core model (LIN).

equations (5.72) to (5.74), or in the outer iterations. A similar problem arises in any type of linearized core control approach.

5.2.6.2. Constant-Load Three-Dimensional Oscillation Damping

The simplified three-dimensional core model was controlled by six independent control elements: one azimuthally symmetric, strong full-length control rod bank (bank 1, Figure 5.3), one full-length control rod bank in each quadrant (banks 2 to 5), and the boron concentration in the coolant. The control problem formulation employed the equilibrium power, xenon, and iodine distributions at 100% power level, since the desired distributions and the weighting factors were the same for all time steps in the present case but were time-varying in Section 5.2.6.3. Constraints were specified on the control variable operating ranges and speeds and on the zonal power densities.

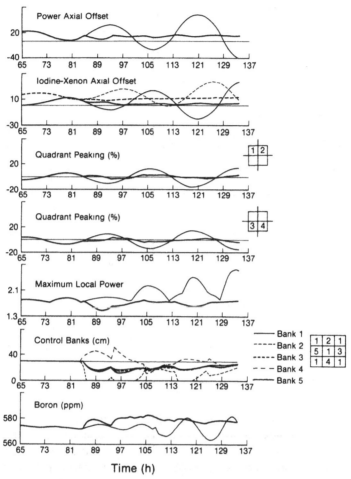

Figure 5.3. Control of large-amplitude three-dimensional xenon oscillation applying a constraint on the maximum local power density (heavy lines = controlled, thin lines = uncontrolled).

The control studies were initiated by perturbing the power distribution with bank 2 from 10 to 14 h of simulated time. The reactivity effects were compensated for by boron. This perturbation triggered both axial and azimuthal (side to side in this rectangular case) oscillations. These oscillations were allowed to grow before the control was started. The uncontrolled growing oscillations are shown by the thin lines in Figure 5.3. The control was initiated at 84 h and at 19 h in Sections 5.2.6.2 and 5.2.6.3, respectively. From those times on, control variable values calculated by MMP were simulated. The control was calculated for about 10 h ahead at a time, without

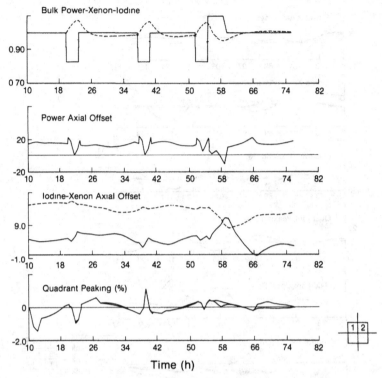

Figure 5.3. (*continued*).

any iterations, and then simulated with the simulator. A new control sequence for the next control period was again calculated, starting from the last simulated state, and so on. The control was thus of the open-loop type for each control period, but feedback information from the real process (simulator) was obtained in the form of the initial state for a new control sequence calculation.

In Figure 5.3, control was started first when both axial and azimuthal oscillations had reached very large amplitudes. The controlled behavior is shown with the heavy lines. Both the axial and azimuthal oscillations are very effectively damped, and the maximum local (zonal) power density is maintained below the maximum allowable value. The azimuthal control does not perturb the axial control, since the method inherently takes into account the overall effect of all the control actions. The azimuthal oscillations in this case are from side to side, from the side of quadrants 1 and 2 to the side of quadrants 3 and 4, and therefore banks 2 and 4 are in practice controlling the azimuthal oscillations. Banks 1, 3, and 5 are in this case seen practically as one bank to control the axial oscillations.

In this case, just the control variable constraints and total power constraint influence the control; the local power constraints do not hit their constraints limit values.

5.2.6.3. Variable-Load Three-Dimensional Control

Core model and control problem formulation in this section are the same as in Section 5.2.6.2, except for the load demand, which is not constant. Only the MMP-controlled behavior, from 19 h onward, is shown in Figure 5.4. The reactor was scheduled to make three load cycles with 5% min^{-1} ramps rates, as indicated by the solid line representing the total power in the uppermost figure. During the third cycle, however, the power level was raised to 110% instead of 100%. It remained at that level for 4 h and was reduced to 100% at a slower ramp rate. The load increases took place at the time of maximum xenon poisoning, which is most demanding for the control system. The control constraints inhibited the use of boron for rapid load changes, and the local power constraints required the zonal power densities to be less than the maximum power density in the core at the 100% power equilibrium state with the control rod banks slightly inserted.

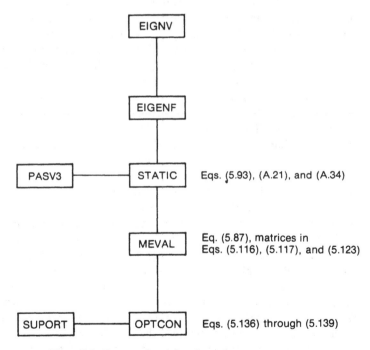

Figure 5.4. Computational flowchart of computer programs.

The MMP control shows a clear anticipatory nature. Specially, before load increases, and while still at low power, preconditioning of the core is quite evident. The power distribution is not strictly controlled to the desired one, but rather the core and the control actuators are brought to a state of preparedness for return to 100% power and subsequent maintenance of the desired power distribution. Therefore, the control banks are withdrawn after the load reduction, inserted before the load increase, and withdrawn during the load increase. The rod insertion before load increases opposes the power distribution swing toward the top of the core resulting from the subsequent rod withdrawal during the load increase.

The power peaking in the upper core half after load increases would be larger without this control action. The control during the reduced power period results also in some excess xenon in the core upper half, an effect that also opposes the upward power distribution swing. The iodine distribution is maintained close to its equilibrium shape in spite of the power distribution changes. After return to 100% power, the core is very close to the desired equilibrium state with respect to the normalized distributions, and practically only boron adjustments are necessary to compensate for bulk xenon. During the whole maneuvering period, the local powers are kept below the permitted values.

The planar power distribution control requires only small, azimuthally nonsymmetric control actions, because the amplitudes of the azimuthal oscillations are small. The quadrant peakings show that these unstable oscillations are also effectively controlled during variable load operation.

The control during the 110% power operation differs from the rest of the transient and is dominated by the local power constraints. The control rod bank movements necessary to reach the 110% power level are so large that the power distribution is strongly swinging upward when the 110% level has been reached. To maintain the power level and to satisfy the local power constraints, bank 1, with the center rod, is inserted, and all the other banks are withdrawn entirely. That effectively reduces the local power in the upper central part of the core while the power in the outer regions of the core is increased with the withdrawal of the other banks. This control action clearly demonstrates the three-dimensional nature of the method and how the hard constraints influence the control. Only a minor overshoot of the local power over the permitted value occurs as a result of the small discrepancy between the process (simulator) behavior and that predicted by the control model. Without the local power constraints, the overshoot would be much larger, as a reference run showed. In the control optimization, the calculated control trajectory moves along the constraint boundary here. The overpower operation with the local power constraint results in power, xenon, and iodine distributions that are far from equilibrium; however, after the overpower operation, the distributions are again controlled toward the equilibrium.

5.2.7. Concluding Remarks

The MMP method is based on control models derived by linearizing a three-dimensional modal diffusion model with respect to xenon redistribution, control actions, and thermal feedbacks. Some nonlinearity is introduced by the use of a nonlinear point model for representing the power, xenon, and iodine bulk levels in the core and by making the control effects power-level dependent. Iterative techniques can be used to further improve the control model performance. The control problem is formulated with an objective function based on desired power, xenon, and iodine distributions, and constraints on the core state and control actions. The control period is divided into time steps, and a constrained multistage optimization problem is formulated. The total power demand is assumed to be known, and the control for the whole control period is solved simultaneously. Smooth control is achieved because the control speeds rather than the control variable values are obtained as the solution. In this procedure, the known load schedule has a feedforward-type effect, and the control results in preconditioning of the core for load changes, especially at low power when some freedom exists in the power distribution. Simulation results indicate that the control models appear to be sufficiently accurate for control purposes. However, further investigations are needed with respect to the amount of spatial detail required. The simulations also clearly demonstrate the anticipatory nature of the control and the effect of the control and state variable constraints on the control.

Important features of the method are as follows:

1. It is three-dimensional.
2. It results in an anticipatory control.
3. It explicitly considers the constraints on the control variables and core state.

On the other hand, the solution is computationally fairly complicated and is obtained in an open-loop form. Furthermore, the operational use requires a fairly detailed estimation of the core state. Current work on the method is concentrated on the questions of the spatial detail needed in the control model and the operational use of the method.

5.3. Xenon Spatial Oscillations in Load-Following: Linear-Quadratic Problem (Ref. 5.3)

In Section 5.2 the problem of controlling the total power and power distribution in a large reactor core has been studied The control problem is solved subject to hard constraints, which can be applied on total power, control variables, and their rate of change, etc. Based on a three-dimensional linearized nodal core model with some slightly nonlinear features, the optimal control problem is solved by quadratic programming.

This section is devoted to solving the problem of control of the xenon spatial oscillations in load-following operations of a nuclear reactor. The model is formulated as a linear-quadratic tracking problem in the context of modern optimal control theory, and the resulting two-point boundary value problem is solved directly by the techniques of initial value methods. The proposed technique is applied to a current-design nuclear PWR.

5.3.1. The Reactor Feedback Model

The reactor core is described by the one-group diffusion model with moderator temperature and xenon-iodine feedbacks as follows:

$$\frac{1}{\nu}\frac{\partial \Phi(r, \theta, x, t)}{\partial t} = \nabla D\nabla\Phi - \Sigma_a\Phi + \nu\Sigma_f\Phi - \sigma_x X\Phi - \Sigma_c\Phi$$

$$- c_b\sigma_B BN_w\Phi - \alpha_m\nu\Sigma_f(T_c - \bar{T}_c)\Phi, \qquad (5.75)$$

$$\frac{\partial I}{\partial t}(r, \theta, z, t) = \gamma_I\Sigma_f s\Phi - \lambda_I I \qquad (5.76)$$

and

$$\frac{\partial X}{\partial t}(r, \theta, z, t) = \gamma_x\Sigma_f\Phi + \lambda_I I - \lambda_x X - \sigma_x X\Phi \qquad (5.77)$$

where, in addition to conventional terminology,

σ_B = average microscopic neutron absorption cross section of ^{10}B
B = boron concentration, ppm
c_b = conversion factor for boron concentration from parts per million to atoms of ^{10}B per molecule of water
N_w = water molecule number density
α_m = moderator temperature reactivity coefficient in $\Delta k/k$ per degrees Celsius
T_c = coolant temperature, °C
$\bar{T}_c$ = reference coolant temperature, °C
Σ_c = control rod absorption cross section

Considering a bare reactor, we give boundary conditions to Φ:

$$\Phi(R, \theta, z, t) = 0, \qquad \Phi(r, \theta, 0, t) = 0, \qquad \Phi(r, \theta, L, t) = 0$$

where radius R and height L include the extrapolated distances. Initial conditions are

$$\Phi(r, \theta, z, 0) = \Phi_0(r, \theta, z)$$

$$I(r, \theta, z, 0) = I_0(r, \theta, z)$$

$$X(r, \theta, z, 0) = X_0(r, \theta, z)$$

We are particularly interested in the case when initial conditions Φ_0, I_0, and X_0 are the steady-state solutions of equations (5.75) to (5.77).

The form of the moderator temperature feedback arises from an approximation of first-order perturbation theory; i.e.,

$$\Delta\rho = \alpha_m(T_c - \bar{T}_c) \simeq \left(\frac{\Phi^+, \Delta\Sigma_a\Phi}{\Phi^+, \nu\Sigma_f\Phi}\right) \simeq \frac{\Delta\Sigma_a}{\nu\Sigma_f}$$

whence

$$\Delta\Sigma_a \simeq \alpha_m\nu\Sigma_f(T_c - \bar{T}_c) \tag{5.78}$$

The control rod absorption is made up of contributions from a full-length control bank and a part-length control bank as follows:

$$\Sigma_x = A_r\Sigma_{c1}\left[\sum_{i=1}^{N}\frac{1}{r}\delta(r - r_i)\delta(\theta - \theta_i)\right]H[z - z_1(t)]$$

$$+ A_r\Sigma_{c2}\left[\sum_{j=1}^{M}\frac{1}{r}\delta(r - r_j)\delta(\theta - \theta_j)\right]$$

$$\times \{H[z - z_2(t)] - H[z - z_2(t) - l]\} \tag{5.79}$$

where

A_r = effective cross-sectional area of a control rod cluster
Σ_{c1} = absorption cross section of a full-length control rod
Σ_{c2} = absorption cross section of a part-length control rod
r_i, θ_i = polar position coordinates of the ith full-length control rod cluster
r_j, θ_j = polar position coordinates of the jth part-length control rod cluster
$z_1(t)$ = position of the end of the full-length control bank
$z_2(t)$ = position of the end of the part-length control bank
l = length of absorbing material in the part-length control rod
N = number of full-length control rod clusters in a bank
M = number of part-length control rod clusters in a bank
δ, H = Dirac delta function and Heaviside function, respectively

Current commercial-sized PWRs are inherently unstable in the axial direction. We consider such a reactor and concentrate on the axial behavior in this section.

We assume that the dependent variables in equations (5.75) to (5.77) are separable in the radial and axial directions for an azimuthally symmetric reactor; i.e.,

$$\Phi(r, \theta, z, t) = J_0(B_r r)\psi(z, t) \tag{5.80}$$

$$I(r, \theta, z, t) = J_0(B_r r)d(z, t) \tag{5.81}$$

and

$$X(r, \theta, z, t) = J_0(B_r r)X(z, t) \tag{5.82}$$

where B_r^2 is the radial "buckling" and J_0 is the zeroth-order Bessel function. Substituting these relations and integrating over the r and θ variables yields

$$\frac{1}{v}\frac{\partial \psi}{\partial t} = D\frac{\partial^2 \psi}{\partial z^2} + [\nu\Sigma_f - \Sigma_a - DB_r^2 + \alpha_m \nu\Sigma_f \bar{T}_c$$

$$- \overline{\alpha_m \nu}\Sigma_f T_c + \alpha_m \omega \nu\Sigma_f T_{in}(t)$$

$$- \bar{\sigma}_x X - \Sigma_c(z, t) - c_b \sigma_B BN_w(z, t)]\psi \tag{5.83}$$

$$\frac{\partial d}{\partial t} = \gamma_I \Sigma_f \psi - \lambda_I d \tag{5.84}$$

$$\frac{\partial X}{\partial t} = \gamma_x \Sigma_f \psi + \lambda_I^d - \lambda_x X - \sigma_x X\psi \tag{5.85}$$

where we have defined

$$\bar{\sigma}_x \triangleq \frac{\bar{J}_0}{\bar{J}_0}\sigma_x, \quad \bar{J}_0 \triangleq \int_0^R rJ_0\, dr, \quad \bar{\bar{J}}_0 \triangleq \int_0^R r^2 J_0\, dr$$

$$\bar{\alpha}_m \triangleq \frac{R^2}{2}\frac{\bar{\bar{J}}_0}{\bar{J}_0^2}\alpha_m, \quad \omega \triangleq \frac{\bar{\alpha}_m}{\alpha_m} - 1 \tag{5.86}$$

and used the relation

$$T_c(z, t) \simeq \frac{2\pi\varepsilon_f \Sigma_f \bar{J}_0}{WC_p}\int_0^z \phi(z', t)\, dz' + T_{in}(t) \tag{5.87}$$

where ε_f = recoverable energy per fission
   W = total coolant mass flow rate
   C_p = heat capacity of coolant
   T_{in} = coolant inlet temperature

The control rod absorption cross section in equation (5.83) resulting from the assumption of separability is

$$\Sigma_c(z, t) \triangleq \frac{A_r \Sigma_{c1}}{2\pi\bar{J}_0}\sum_{i=1}^N J_0(B_r r_i)H[z - z_i(t)]$$

$$+ \frac{A_r \Sigma_{c2}}{2\pi\bar{J}_0}\sum_{j=1}^M J_0(B_r r_j)\{H[z - z_2(t)] - H[z - z_2(t) - l]\} \tag{5.88}$$

Since we are interested in control of perturbations from some steady state, we first find the steady-state solution around which the nonlinear system of equations (5.83) to (5.85) is to be linearized. The steady-state form of equation (5.83) is obtained by setting the time derivative to zero and using the steady-state versions of equations (5.84), (5.85), (5.87), and (5.88). The resulting steady-state neutron balance with xenon, iodine, and moderator temperature feedback and with boron and control rod control is

$$0 = D\frac{d^2\psi_0}{dz^2} + (\nu\Sigma_f - \Sigma_a - DB_r^2 - \Sigma_{c0} + \alpha_m\nu\Sigma_f\bar{T}_c + \alpha_m\omega\nu\Sigma_f T_{in,0})$$

$$- \frac{\bar{\sigma}_x(\gamma_i + \gamma_x)\Sigma_f\psi_0^2}{\lambda_x + \bar{\sigma}_x\psi_0} - c_b\sigma_B B_0 y_0 N_{w0}\left[1 - \frac{2\pi\varepsilon_f\Sigma_f\bar{J}_0\beta}{WC_p}\int_0^z \psi_0(z')\,dz'\right]$$

$$- \bar{\alpha}_m\nu\Sigma_f\psi_0\left[\frac{2\pi\varepsilon_f\Sigma_f\bar{J}_0}{WC_p}\int_0^z \psi_0(z')\,dz' + T_{in,0}\right] \tag{5.89}$$

where we used the coolant volumetric coefficient of thermal expansion β:

$$\frac{dN_w}{dT_c} = -\beta N_w \tag{5.90}$$

Using dimensionless variables

$$x \triangleq \frac{z}{L} \tag{5.91}$$

and

$$u \triangleq \frac{\bar{\sigma}_x}{\lambda_x}\psi_0 \tag{5.92}$$

where L is the axial length of the core, we can write the steady-state equation in operator form as

$$Lu + \lambda u = g(x, u), \qquad x \in [0, 1]$$
$$u(0) = u(1) = 0 \tag{5.93}$$

where

$$L = \frac{d^2}{dx^2} - a_0(x) \tag{5.94}$$

$$a_0(x) = \frac{L^2}{D}[DB_r^2 + \Sigma_{c0}(x) + \alpha_m\nu\Sigma_f(T_{in,0} - \bar{T}_c) + c_b\sigma_B B_0 N_{w0}(0)] \tag{5.95}$$

$$\lambda = \frac{L^2}{D}(\nu\Sigma_f - \Sigma_a) \tag{5.96}$$

$$g(x, u) = \hat{A}\frac{u^2}{1 + u} + \hat{B}u \int_0^x u(x') \, dx' \qquad (5.97)$$

$$\hat{A} = \frac{L^2}{D}(\gamma_I + \gamma_x)\Sigma_f \qquad (5.98)$$

$$\hat{B} = \frac{L^3}{D}\frac{\lambda_x}{\bar{\sigma}_x}\frac{2\pi\varepsilon_f\Sigma_f\bar{J}_0}{WC_p}[\bar{\alpha}_m\nu\Sigma_f - c_b\sigma_B B_0 N_{w0}(0)\beta] \qquad (5.99)$$

Equation (5.93) belongs to a class of nonlinear eigenvalue problems. Existence and uniqueness of solutions to this class of problems are the subjects of bifurcation theory. Applications of the theory to nuclear reactor problems have previously been made with feedback.

5.3.2. The Steady-State Solution

By using bifurcation theory, it is shown in Appendix A that the non-linear eigenvalue problem (5.93) with $B > 0$ has a unique positive solution for $\lambda > \lambda_0$, where λ_0 is called the *bifurcation point*. In our model, $\hat{B} > 0$ means that physically the negative temperature feedback of the moderator overweighs the positive temperature feedback of the boron concentration.

The value λ_0 corresponds to the critical configuration of the multiregion reactor with nonlinear feedbacks removed. Appendix A also shows that the unique positive solution can be obtained by an iteration scheme based on monotone operators. A convenient starting solution for the iteration is the scaled fundamental eigenfunction $\varepsilon\psi_0$ of the linearized equation (5.19) around $u_0(x) = 0$. Details of finding λ_0 and ψ_0 are in Reference 5.20.

5.3.3. Linearized Equations

Having found the steady-state solution, we now return to equations (5.83) to (5.85) and linearize them around the steady state. Neglecting second- and higher-order deviations, we get

$$\frac{1}{\nu}\frac{\partial \delta\psi}{\partial t} = D\frac{\partial^2 \delta\psi}{\partial z^2} + (\nu\Sigma_f - \Sigma_a - DB_r^2 + \alpha_m\nu\Sigma_f\bar{T}_c)\delta\psi - \bar{\sigma}_x\chi_0\delta\psi - \bar{\sigma}_x\psi_0\delta_x$$

$$- \Sigma_{c0}\delta_x - \psi_0\delta\Sigma_c - c_b\sigma_B B_0 N_{w0}\delta\psi - c_b\sigma_B B_0\psi_0\delta N_w$$

$$- c_b\sigma_B N_{w0}\psi_0\delta B - \bar{\alpha}_m V\Sigma_f T_{c0}\delta\psi - \bar{\alpha}mV\Sigma_f\psi_0\delta T_c$$

$$+ \alpha_m\omega\nu\Sigma_f T_{in,0}\delta\psi + \alpha_m\omega\nu\Sigma_f\psi_0\delta T_{in} \qquad (5.100)$$

$$\frac{\partial \delta_d}{\partial t} = \gamma_I\Sigma_f\psi - \lambda_I\delta \qquad (5.101)$$

and

$$\frac{\partial \delta_\chi}{\delta t} = \gamma_\chi \Sigma_f \delta\psi + \lambda_I \delta - \lambda_\chi \delta_\chi - \bar{\sigma}_\chi \chi_0 \delta\psi - \bar{\sigma}\psi_0 \delta_\chi \qquad (5.102)$$

where

$$\delta\Sigma_c = -\frac{A_r \Sigma_{c1}}{2\pi \bar{J}_0} \sum_{i=1}^{N} J_0(B_r r_i)\underline{\delta}(z - z_{10})\delta z_1(t)$$

$$-\frac{A_r \Sigma_{c2}}{2\pi \bar{J}_0} \sum_{j=1}^{M} J_0(B_r r_j)[\underline{\delta}(z - z_{20}) - \underline{\delta}(z - z_{20} - l)]\delta z_2(t) \qquad (5.103)$$

$$\delta T_c \simeq \frac{2\pi \varepsilon f \Sigma_f J_0}{WC_p} \int_0^z \delta\psi(z't) \, dz' + \delta T_{in}(t) \qquad (5.104)$$

$$\delta N_w \simeq -\beta N_w \delta T_c = -\beta N_{w0}\delta T_c$$

$$= -\beta N_{w0} \frac{2\pi \varepsilon_f \Sigma_f \bar{J}_0}{WC_p} \int_0^z \delta\psi(z', t) \, dz' - \beta N_{w0}\delta T_{in}(t) \qquad (5.105)$$

where $\underline{\delta}$ is the Dirac delta function, and δ is the deviation. The boundary conditions are

$$\delta\psi(0, t) = 0 \qquad (5.106)$$

and

$$\delta\psi(L, T) = 0 \qquad (5.107)$$

The initial conditions are

$$\delta\psi(z, 0) = 0 \qquad (5.108)$$

$$\delta d(z, 0) = 0 \qquad (5.109)$$

and

$$\delta\chi(z, 0) = 0 \qquad (5.110)$$

We now expand $\delta\psi(z, t)$, $\delta d(z, t)$, and $\delta\chi(z, t)$ in eigenfunctions of the Helmholtz equation $\psi_i(z)$; i.e.,

$$\delta\psi(z, t) = \sum_{i=1}^{N_a} a_i(t)\psi_i(z) \qquad (5.111)$$

$$\delta d(z, t) = \sum_{i=1}^{N_b} b_i(t)\psi_i(z) \qquad (5.112)$$

and

$$\delta\chi(z, t) = \sum_{i=1}^{N_c} c_i(t)\psi_i(z) \qquad (5.113)$$

where $\psi_i(z)$ is the eigenfunction of

$$\frac{d^2\psi_i}{dz^2} + k_i^2\psi_i = 0, \qquad \psi_i(0) = \psi_i(L) = 0 \qquad (5.114)$$

that is,

$$\psi_i(z) = \sin\frac{i\pi z}{L}, \qquad i = 1, 2, \ldots \qquad (5.115)$$

We then substitute these into the linearized equations and form inner products with ψ_j; then for the two-term expansion ($N_a = N_b = N_c = 2$), we obtain

$$\frac{1}{v}\dot{\zeta} = A\zeta + B\xi + Cu, \qquad \zeta(t_0) = 0 \qquad (5.116)$$

and

$$\dot{\xi} = D\xi + E\zeta, \qquad \xi(t_0) = 0 \qquad (5.117)$$

where the state variables are

$$\zeta(t) \triangleq \begin{bmatrix} a_1(t) \\ a_2(t) \end{bmatrix}, \qquad \xi(t) \triangleq \begin{bmatrix} b_1(t) \\ b_2(t) \\ c_1(t) \\ c_2(t) \end{bmatrix} \qquad (5.118)$$

and the control variables are

$$u(t) \triangleq \begin{bmatrix} \delta z_1(t) \\ \delta z_2(t) \\ \delta B(t) \\ \delta T_{in}(t) \end{bmatrix} \qquad (5.119)$$

Evaluations of the elements of matrices A, B, C, D, and E are given in Reference 5.20.

5.3.4. Performance Index and the Optimal Control Problem (Refs. 5.22–5.25)

We choose as the performance index a quadratic functional

$$J = \frac{1}{2}\int_{t_0}^{t_f} dt \left[\int_V dV \left(\varepsilon_f \Sigma_f \Phi - \varepsilon_f \Sigma_f \Phi_d P\right)^2 + u^T Ru \right] \qquad (5.120)$$

where V = reactor volume

Φ_d = desired flux spatial distribution

$P(t)$ = desired fraction of the total reactor power to be followed in time $t_0 \to t_f$

$R = \text{diag}\{\beta_1, \beta_2, \beta_3, \beta_4\}$

β_i = weighting parameters for each control variable, $i = 1, 2, 3, 4$

Using equation (5.80) and the two-term expansion in equation (5.11), we get

$$J = \tfrac{1}{2} \int_{t_0}^{t_f} dt \, [2\pi(\varepsilon_f \Sigma_f)^2 \bar{J}_0 \int_0^L dz \, (\psi_0 + a_1\psi_1 + a_2\psi_2 - \psi_d P)^2 + u^T R u]$$

(5.121)

where we assume that

$$\Phi_d = J_0(B_r r)\psi_d(z)$$

(5.122)

Let

$$\Lambda = 2\pi(\varepsilon_f \Sigma_f)^2 \bar{J}_0 \int_0^L dz \, (\psi_0 + a_1\psi_1 + a_2\psi_2 - \psi_d P)^2 + u^T R u$$

and

$$K = 2\pi(\varepsilon_f \Sigma_f)^2 \bar{J}_0$$

$$\psi = \begin{bmatrix} \psi_1 \\ \psi_2 \end{bmatrix}$$

then

$$\Lambda = K \int_0^L dz \, (\psi^T \zeta + \psi_0 - \psi_d P)^T (\psi^T \zeta + \psi_0 - \psi_d P) + u^T R u$$

$$= K \int_0^L dz \, [\zeta^T \psi \psi^T \zeta + 2(\psi_0 - \psi_d P)\psi^T \zeta + (\psi_0 - \psi_d P)^2] + u^T R u$$

or

$$\Lambda = \zeta^T Q_1 \zeta + 2(Q_2 - PQ_3)\zeta + Q_4 + u^T R u$$

(5.123)

where

$$Q_1 \triangleq K \int_0^L dz \, (\psi \psi^T)$$

$$Q_2 \triangleq K \int_0^L dz \, (\psi_0 \psi^T)$$

$$Q_3 \triangleq K \int_0^L dz \, (\psi_d \psi^T)$$

$$Q_4 \triangleq K \int_0^L dz \, (\psi_0 - \psi_d P)^2$$

Noting that Q_1 is nonsingular and symmetric, we have

$$\Lambda = [\zeta + Q_1^{-1}(Q_2^T - PQ_3^T)]Q_1[\zeta + Q_1^{-1}(Q_2^T - PQ_3^T)]$$
$$- (Q_2 - PQ_3)Q_1^{-1}(Q_2^T - PQ_3^T) + Q_4 + u^T Ru \qquad (5.124)$$

The second and third terms in equation (5.124) are specified functions of time and do not involve state or control variables, so it is irrelevant to include them in the performance index. Therefore, the final form of our performance index becomes

$$J = \tfrac{1}{2} \int_{t_0}^{t_f} [(\zeta - \zeta_d)^T Q_1(\zeta - \zeta_d) + u^T Ru]\, dt \qquad (5.125)$$

where

$$\zeta_d(t) \triangleq -Q_1^{-1}[Q_2^T - P(t)Q_3^T] \qquad (5.126)$$

The control problem of equations (5.116), (5.117), and (5.125) is now in the form of an optimal tracking problem.

An analysis using the singular perturbation theory (Ref. 5.21) indicates that, for the time scales characteristic of the dynamic feedback problem treated here, equation (5.116) may be approximated by the reduced form

$$0 = A\zeta + B\xi + Cu \qquad (5.127)$$

which is analogous to the "prompt jump approximation" in reactor point kinetics (Ref. 5.22).

Solving equation (5.127) for ζ, we get

$$\zeta = B''\xi + C''u \qquad (5.128)$$

where

$$B'' = A^{-1}B, \qquad C'' = A^{-1}C \qquad (5.129)$$

Equation (5.117) becomes

$$\dot{\xi} = D'\xi + C'u \qquad (5.130)$$

where

$$D' = D + EB'', \qquad C' = EC'' \qquad (5.131)$$

Applying classical optimal control theory (Ref. 5.21), we write the Hamiltonian as

$$H = \tfrac{1}{2}(\zeta - \zeta_d)^T Q_1(\zeta - \zeta_d) + \tfrac{1}{2}u^T Ru + \lambda^T(D'\xi + C'u)$$
$$= \tfrac{1}{2}(B''\xi + C''u - \zeta_d)^T Q_1(B''\xi + C''u - \zeta_d)$$
$$+ \tfrac{1}{2}u^T Ru + \lambda^T(D'\xi + C'u) \qquad (5.132)$$

A necessary condition for optimality is that

$$O = \frac{\partial H}{\partial u} = C''^T Q_1 (B''\xi + C''u - \zeta_d) + Ru + C'^T\lambda$$

Solving for the control vector u gives

$$u = -R'^{-1}(C''^T Q_1 B''\xi - C''^T Q_1 \zeta_d + C'^T\lambda) \qquad (5.133)$$

where

$$R' \triangleq R + C''^T Q_1 C''$$

The adjoint state equation is

$$\dot{\lambda} = \frac{\partial H}{\partial \xi} = -B''^T Q_1 (B''\xi + C''u - \zeta_d) - D'^T\lambda, \qquad \lambda(t_f) = 0$$

Using equation (5.133) for u, we get

$$\dot{\lambda} = -B''^T Q_1 B''\xi + B''^T Q_1 C'' R'^{-1}(C''^T Q_1 B''\xi - C''^T Q_1 \zeta_d + C'^T\lambda)$$
$$+ B''^T Q_1 \zeta_d - D'^T\lambda, \qquad \lambda(t_f) = 0$$

or

$$\dot{\lambda} = (-B''^T Q_1 B'' + B''^T Q_1 C'' R'^{-1} C''^T Q_1 B'')\xi + (B''^T Q_1 C'' R'^{-1} C'^T - D'^T)\lambda$$
$$+ (-B''^T Q_1 C'' R'^{-1} C''^T Q_1 + B''^T Q_1)\zeta_d, \qquad \lambda(t_f) = 0 \qquad (5.134)$$

The state equation is

$$\dot{\xi} = \frac{\partial H}{\partial \lambda} = D'\xi + C'u, \qquad \xi(t_0) = 0$$

Using equation (5.133) for u gives

$$\dot{\xi} = D'\xi - C'R'^{-1}(C''^T Q_1 B''\xi - C''^T Q_1 \zeta_d + C'^T\lambda)$$

or

$$\dot{\xi} = (D' - C'R'^{-1}C''^T Q_1 B'')\xi - C'R'^{-1}C'^T\lambda + C'R'^{-1}C''^T Q_1 \zeta_d,$$
$$\xi(t_0) = 0 \qquad (5.135)$$

Rewriting equations (5.134) and (5.135) in a more compact form, we note that we have a linear TPBVP of dimension 8:

$$\dot{\xi} = A_1\xi + B_1\lambda + C_1\zeta_d, \qquad \xi(t_0) = 0 \qquad (5.136)$$
$$\dot{\lambda} = A_2\xi + B_2\lambda + C_2\zeta_d, \qquad \lambda(t_f) = 0 \qquad (5.137)$$

where

$$A_1 \triangleq D' - C'R'^{-1}C''^T Q_1 B''$$

$$B_1 \triangleq -C'R'^{-1}C'^T$$

$$C_1 \triangleq C'R'^{-1}C''^T Q_1$$

$$A_2 \triangleq -B''^T Q_1 B'' + B''^T Q_1 C'' R'^{-1} C''^T Q_1 B''$$

$$B_2 \triangleq B''^T Q_1 C'' R'^{-1} C'^T - D'^T$$

$$C_2 \triangleq -B''^T Q_1 C'' R'^{-1} C''^T Q_1 + B''^T Q_1$$

Rewriting equation (5.133) for the control vector u and equation (5.128), we have

$$u = A_3 \xi + B_3 \lambda + C_3 \zeta_d \qquad (5.138)$$

where

$$A_3 \triangleq -R^{-1}C''^T Q_1 B''$$

$$B_3 \triangleq -R'^{-1}C'^T$$

$$C_3 \triangleq R'^{-1}C''^T Q_1$$

and

$$\zeta = B''\xi + C''u \qquad (5.139)$$

Equations (5.136) through (5.139) constitute the basic relations for the optimal tracking problem defined above. The TPBVP represented by equations (5.136) and (5.137) is solved by using the computer program SUPORT (Ref. 5.23). Appendix B describes the solution method.

5.3.5. Calculation Scheme and Results

The reactor parameters used in the calculations are given in Table 5.1. These are representative of current designs of commercial PWR.

The computational flowchart identifying steps in the computations leading to and including the optimal control program is shown in Figure 5.4. Two existing programs for TPBVPs—PASVA3, a revision of PASVAR, and SUPORT—were used in the calculations, and other computer programs were generated.

The scaled fundamental eigenfunction from EIGNF is taken as the starting solution. From the iterated steady-state flux distribution, the computer program MEVAL is used to calculate the coolant temperature and density. These results are shown in Figures 5.5 and 5.6. MEVAL also

Table 5.1. Design Data for a PWR

Thermal output, MW (thermal)	3400
Active core height, cm	370
Equivalent core diameter, cm	340
System pressure, bar	155
Coolant mass flow rate, kg/h	56.5×10^6
Coolant inlet temperature, °C	300
Coolant inlet density, g/cm^3	0.72
Coolant heat capacity, J/kg · °C	6.06×10^3
Coolant thermal expansion coefficient, °C^{-1}	3×10^{-3}
Coolant temperature reactivity coefficient, $\Delta\rho/$°C	4.73×10^{-4}
Absorption cross section of ^{10}B, b	2.207×10^3
Absorption cross section of ^{135}Xe, b	2.36×10^6
Fractional fission yield of ^{135}Xe	0.00228
Fractional fission yield of ^{135}I	0.06386
Decay constant of ^{135}Xe, h^{-1}	0.0753
Decay constant of ^{135}I, h^{-1}	0.1035

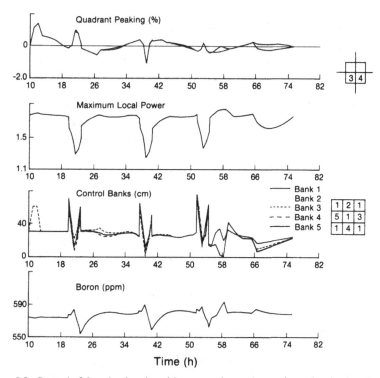

Figure 5.5. Control of three load cycles with a constraint on the maximum local power density.

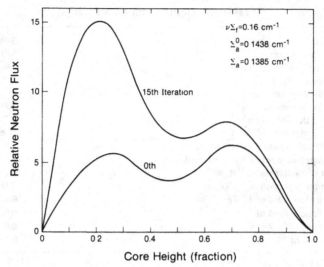

Figure 5.6. Starting solution of the 15th iteration in steady-state calculation (the flux has units of 1.42×10^{13} N/cm$^2 \cdot$ sec).

performs evaluations of the elements of matrices A, B, C, D, E, Q_1, Q_2, Q_3, and the necessary scaling.

The computer program OPTCON performs the necessary matrix operations and solves the linear TPBVP, equations (5.136) and (5.137), by calling SUPORT.

We show selected numerical results of the optimal solutions in Figures 5.7 through 5.12. Several observations may be noted. Each case studied shows that the optimal solution closely follows the desired load demand

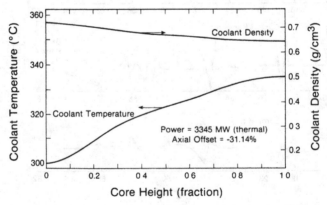

Figure 5.7. Steady-state coolant temperature and density distribution.

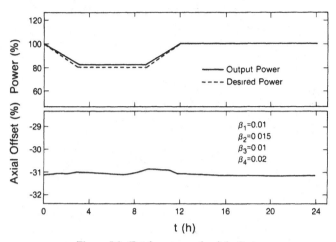

Figure 5.8. Total power and axial offset.

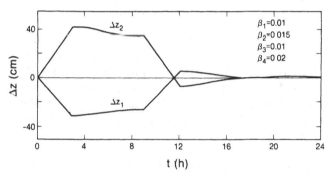

Figure 5.9. Full- and part-length control rod bank deviation: full length $= \Delta z_1$; part length $= \Delta z_2$.

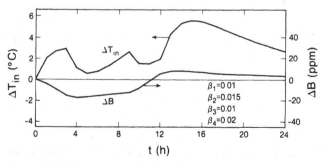

Figure 5.10. Boron concentration and inlet temperature deviations.

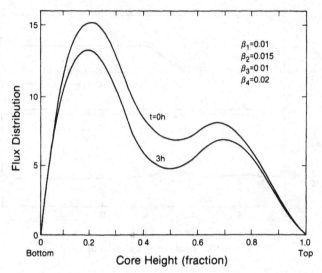

Figure 5.11. Flux distribution in core height and time (the flux has units of 1.42×10^{13} N/cm$^2 \cdot$ sec).

(82% for 80% target) and maintains the desired power distribution (essentially constant axial offset) while expending a small control effort.

A full-length control rod bank and the part-length control rod bank move in opposite directions, and they accomplish most of the control to maintain the desired power distribution during power-level changes. Several

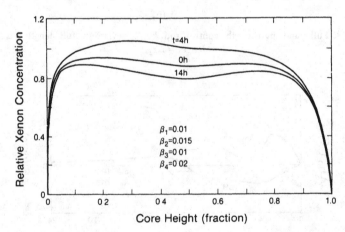

Figure 5.12. Xenon distribution in core height and time (xenon has units of 2.03×10^{15} atom/cm^3).

hours after achieving full power, both control rod banks return to their initial steady-state positions.

After power-level changes, the coolant inlet temperature and boron concentration changes accomplish most of the control to compensate for reactivity change due to the delayed xenon feedback until the end of the control period.

Since we utilize the current state-of-the-art in initial value methods for the TPBVP, optimal solutions are found with a minimum amount of computer time. We believe that it is important to obtain an accurate steady-state solution around which linearization is to take place, and that the optimal open-loop control strategy strongly depends on the initial steady-state power distribution. It is therefore recommended that design codes be fully utilized to generate accurate and realistic initial power distributions. There should be no theoretical difficulty in this model for the extension to closed-loop control because the model is linear and quadratic without "hard" constraints. However, it will require an efficient core estimator. The limitations of a linear analysis for what is essentially a nonlinear feedback problem could be offset by precalculating steady-state power distributions at several power levels and then using a "time-synthesis" technique (Ref. 5.24).

5.3.6. Concluding Remarks

We analyzed the problem of spatial control of a nuclear PWR designed to follow variations of the electrical load. The reactor is subject to feedbacks from xenon poisoning and from moderator temperature and density effects. It is controlled by the action of part-length and full-length control rods, by the uniform variation of boron concentration, and by changes in the inlet coolant temperature. The open-loop control obtained is optimal in the sense of minimizing a quadratic functional of the power distribution, which expresses the square of the deviation of the actual power from the desired power and the effort required in control. A quadratic "cost functional" is chosen for minimization in order to utilize well-established results of linear systems theory.

Unique features of the present analysis are

1. Inclusion of part-length control rods and the inlet coolant temperature in the control vector
2. An accurate steady-state solution (around which the problem is linearized) that is obtained via nonlinear bifurcation analysis
3. The use of singular perturbation methods for the "stiff" and high-order TPBVP resulting from the application of variational principles to minimize the performance index

5.4. Multilevel Methods (Ref. 5.4)

In Section 5.3, we discussed the problem of controlling the xenon spatial solution in load-following in a large-core reactor. The model is formulated as a linear-quadratic tracking problem, and the theorems of modern optimal control are applied. The resulting optimal equations are formed as a two-point boundary value problem; these equations are solved directly by the technique of initial value methods.

In this section we discuss another multilevel method used to control a BWR with practical operational constraints and thermal limits. Due to the very large size of the problem, a decomposition is made using a hierarchical control technique. The optimization of the resulting subproblems is performed with the feasible direction method.

5.4.1. Meanings of Multilevel Control

The optimal control analysis of a modern BWR is a high-dimensional or "large-scale" problem that involves considerable computational effort. One way of handling this complexity is to treat the overall system as a collection of smaller subsystems that are interconnected by certain parameters to meet overall objectives and constraints. In a multilevel or hierarchical mode, the interconnection parameters are manipulated by a subsystem at a given level that controls or "coordinates" the subsystems at the level just below it. The local subsystems are treated individually on one level while the coordinator operates on the higher level, providing the multilevel structure that gives the process its name. In our analysis, the structure is defined by a discretization of the continuous time control problem into a sequence of interconnected static subsystems.

Consider a dynamic optimal control problem: find

$$\min J = \int_{t_0}^{t_f} j[P(t), U(t)] \, dt + \theta[X(t_f)] \tag{5.140}$$

subject to

$$\dot{X}(t) = f[X(t), U(t), P(t)] \tag{5.141}$$

$$g[X(t), U(t), \psi(t), P(t)] = 0 \tag{5.142}$$

$$h[\psi(t), U(t), P(t)] \le 0 \tag{5.143}$$

and

$$X(t_0) = A \tag{5.144}$$

where

$X(t)$ = n-dimensional state vectors
$U(t)$ = r-dimensional control vectors
$\psi(t)$ = m-dimensional vector relating equality and inequality constraints
$P(t)$ = output of the system
$\theta(t)$ = "penality" on the final state vector X
$f(t)$ = n-dimensional nonlinear continuous dynamic equations
$g(t)$ = m-dimensional nonlinear continuous static vector equations
$h(t)$ = s-dimensional nonlinear continuous constraint vector equations
A = initial vector X

The object is to find the $U(t)$ that minimizes the objective function J and satisfies both the static condition g and constraint h. Equation (5.142), $g = 0$, is a set of equality constraints that connects the variable ψ with the variables X, U, and the output P. The variable ψ is then used to define the inequality constraint H. The penality term θ forces the final state vector X to approach the initial state condition. The usual method of defining the objective function is a periodic problem.

In the following, we explain how to transfer the system equations (5.140) through (5.144) to a multilevel optimal control problem. The equation system is first discretized. The finite time interval (t_0, t_f) is divided into N subintervals with length Δt_k in the kth subinterval. If the variation of the dynamic equation (5.141) is small during time interval Δt_k, we can write

$$X_k = \int_{t_{k-1}}^{t_k} f[X(t), U(t), P(t)]\, dt \simeq F_k(X_{k-1}, U_k, P_k) \qquad (5.145)$$

where subscript k represents the kth subinterval. Equation (5.145) indicates that the state vector X at time t_k is a function of the state vector X at time t_{k-1}, the control vector, and the output at time t_k. The objective function takes the form

$$J = \sum_{k=1}^{N} j_k(P_k, U_k)\Delta t_k + \theta(X_N) \qquad (5.146)$$

$$= \sum_{k=1}^{N} J_k(P_k, U_k) + \theta(X_N) \qquad (5.147)$$

Equations (5.142) and (5.144) also must be satisfied at each subinterval.

The sequences of states are related one to the other, and we define an interconnection parameter Π_k by

$$\Pi_k = X_{k-1} \qquad (5.148)$$

relating the state at step k to that at step $k - 1$. With the aid of equation

(5.148), the problem can be written in discrete form as

$$\min J = \sum_{k=1}^{N} J_k(P_k, U_k) + \theta(X_N) \qquad (5.149)$$

subject to

$$X_k = F_k(\Pi_k, U_k, P_k) \qquad (5.150)$$

$$G_k(X_k, \psi_k, U_k, P_k) = 0 \qquad (5.151)$$

$$H_k(\psi_k, U_k, P_k) \leq 0 \qquad (5.152)$$

$$\Pi_k = X_{k-1} \qquad (5.153)$$

and

$$\Pi_1 = A \qquad (5.154)$$

for $k = 1, \ldots, N$. This discretization converts the dynamic system into a set of N static subsystems that are arranged in a serial structure as shown in Figure 5.13. This characteristic is used to decompose the problem.

The Lagrangian for the problem can be defined as

$$(P_k, X_k \Pi_k, U_k, \psi_k, \beta_k, \rho_k, \lambda_k, \gamma_k) = \sum_{k=1}^{N} J_k(P_k, U_k) + \theta(X_N)$$

$$+ \sum_{k=1}^{N} \beta_k^T \cdot [X_k - F_k(\Pi_k, U_k, P_k)]$$

$$+ \sum_{k=1}^{N} \rho_k^T \cdot (X_{k-1} - \Pi_k)$$

$$+ \sum_{k=1}^{N} \lambda_k^T \cdot G_k(X_k, U_k, \psi_k, P_k)$$

$$+ \sum_{k=1}^{N} \gamma_k^T \cdot H_k(\psi_k, U_k, P_k)$$

$$(5.155)$$

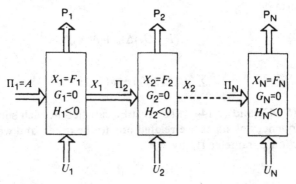

Figure 5.13. Illustration of a set of N sequential subsystems.

where β_k, ρ_k are Lagrange multiplier vectors of dimension n, and λ_k is also an m-dimensional Lagrange vector. They are introduced to take into account the equality constraints. Furthermore, γ_k is a Kuhn–Tucker multiplier vector that takes into account inequality constraints, and it is of dimension s. The component of γ_k is nonzero only if the corresponding constraint is active; by active, we mean that the inequality constraint $h_k^i(\psi_k, U_k, P_k) \leq 0$ at a point (ψ_k, U_k, P_k) is equal to zero.

The optimality conditions are that the Lagrangian is stationary with respect to variations in its arguments and are as follows:

$$\frac{\partial L}{\partial P_k} = \frac{\partial J_k}{\partial P_k} - \left(\frac{\partial F_k}{\partial P_k}\right)^{\mathrm{T}} \cdot \beta_k + \left(\frac{\partial G_k}{\partial P_k}\right)^{\mathrm{T}} \cdot \lambda_k + \left(\frac{\partial H_k}{\partial P_k}\right)^{\mathrm{T}} \cdot \gamma_k = 0 \qquad (5.156)$$

$$\frac{\partial L}{\partial X_k} = \frac{\partial \theta}{\partial X_k} + \beta_k + \rho_{k+1} + \left(\frac{\partial G_k}{\partial X_k}\right)^{\mathrm{T}} \cdot \lambda_m = 0 \qquad (5.157)$$

$$\frac{\partial L}{\partial \Pi_k} = -\left(\frac{\partial F_k}{\partial \Pi_k}\right)^{\mathrm{T}} \cdot \beta_k - \rho_k = 0 \qquad (5.158)$$

$$\frac{\partial L}{\partial U_k} = \frac{\partial J_k}{\partial U_k} - \left(\frac{\partial F_k}{\partial U_k}\right)^{\mathrm{T}} \cdot \beta_k + \left(\frac{\partial H_k}{\partial U_k}\right)^{\mathrm{T}} \cdot \gamma_k + \left(\frac{\partial G_k}{\partial U_k}\right)^{\mathrm{T}} \cdot \lambda_k = 0 \qquad (5.159)$$

$$\frac{\partial L}{\partial \psi_k} = \left(\frac{\partial G_k}{\partial \psi_k}\right)^{\mathrm{T}} \cdot \lambda_k + \left(\frac{\partial H_k}{\partial \psi_k}\right)^{\mathrm{T}} \cdot \gamma_k = 0 \qquad (5.160)$$

$$\frac{\partial L}{\partial \beta_k} = X_k - F_k = 0 \qquad (5.161)$$

$$\frac{\partial L}{\partial \rho_k} = X_{k-1} - \Pi_k = 0 \qquad (5.162)$$

$$\frac{\partial L}{\partial \lambda_k} = G_k = 0 \qquad (5.163)$$

and

$$\frac{\partial L}{\partial \gamma_k} = \gamma_k^{\mathrm{T}} \cdot H_k = 0 \qquad (5.164)$$

for $k = 1, \ldots, N$.

The values $\partial \theta / \partial X_k$ and $\partial J_k / \partial U_k$ are partial derivatives of a scalar with respect to a vector. The matrices $\partial G_k / \partial X_k$, $\partial G_k / \partial U_k$, $\partial F_k / \partial \Pi_k$, $\partial F_k / \partial U_k$, $\partial G_k / \partial \psi_k$, $\partial H_k / \partial U_k$, and $\partial H_k / \partial \psi_k$ represent the partial derivatives of a vector of functions with respect to a vector of variables.

When m, r, n, and s are large, it is difficult to solve directly the nonlinear system given by equations (5.156) through (5.164). We treat the problem here as a multilevel large-scale problem and use decomposition-coordination methods. There are number of approaches; we have chosen the nonfeasible or goal coordination method for the interaction of subsystems and the feasible direction method for the optimization of the subproblems.

The nonfeasible or goal coordination method is characterized by fixing the parameter ρ_k ($k = 1$ to N) on the second level and therefore solving the problems at the first level.

Equation (5.155) is rewritten as

$$L = \sum_{k=1}^{N} [J_k(P_k, U_k) + \beta_k^{\mathrm{T}} \cdot (X_k - F_k) - \rho_{k \cdot \Pi_k}^{\mathrm{T}}$$

$$+ \rho_{k+1}^{\mathrm{T}} \cdot X_k + \lambda_k^{\mathrm{T}} \cdot G_k + \gamma_k^{\mathrm{T}} \cdot H_k] + \theta \qquad (5.165)$$

$$= \sum_{k=1}^{N} L_k(\alpha_k, \delta) + \theta \qquad (5.166)$$

where $\alpha_k^{\mathrm{T}} = (P_k, X_k^{\mathrm{T}}, \Pi_k^{\mathrm{T}}, U_k^{\mathrm{T}}, \psi_k^{\mathrm{T}}, \beta_k^{\mathrm{T}}, \lambda_k^{\mathrm{T}}, \gamma_k^{\mathrm{T}})$ is determined on the first level and $\delta = \rho$ is fixed at the second level.

Equation (5.166) takes a separable form, and each subproblem on the first level is defined by the Lagrangian L_k. In this case, the optimality condition equation (5.162) is satisfied on the second level and the others on the first level.

The first-level subproblems are as follows:

Subproblem 1

$$\min J_1(P_1, U_1) + \rho_2^T \cdot X_1 \qquad (5.167)$$

subject to

$$X_1 = F_1(\Pi_1, U_1, P_1) \qquad (5.168)$$

given

$$\rho_2, G_1(X_1, U_1, \psi_1, P_1) = 0 \qquad (5.169)$$

$$H_1(\psi_1, U_1, P_1) \le 0 \qquad (5.170)$$

and

$$\Pi_1 = A \qquad (5.171)$$

Subproblem k

$$\min J_k(P_k, U_k) + \rho_{k+1}^T \cdot X_k - \rho_k^T \cdot \Pi_k \qquad (5.172)$$

subject to

$$X_k = F_k(\Pi_k, U_k, P_k) \qquad (5.173)$$

given

$$G_k(X_k, U_k, \psi_k, P_k) = 0 \tag{5.174}$$

$$\rho_k, \rho_{k+1}, H_k(\psi_k, U_k, P_k) \leq 0 \tag{5.175}$$

for $k = 2, \ldots, N - 1$

Subproblem N

$$\min J_N(P_N, U_N) + \theta(X_N) - \rho_N^T \cdot \Pi_N \; . \tag{5.176}$$

subject to

$$X_N = F_N(\Pi_N, U_N, P_N) \tag{5.177}$$

given

$$\rho_N, G_N(X_N, U_N, \psi_N, P_N) = 0 \tag{5.178}$$

$$H_N(\psi_N, U_N, P_N) \leq 0 \tag{5.179}$$

The optimal solution of subproblem N, defined by equations (5.176) through (5.179), yields the unbounded Π_N if the penalty term θ is not included. Thus, it is necessary to include the penalty term in equation (5.176), and this is accomplished by allowing θ to impose a limitation on X_N and also on Π_N through equation (5.177).

The modification of the objective function is

$$\rho_2^T \cdot X_1 + \sum_{k=2}^{N-1} (\rho_{k+1}^T \cdot X_k - \rho_k^T \cdot \Pi_k) - \rho_N^T \cdot \Pi_N \tag{5.180}$$

It is zero if $\Pi_k = X_{k-1}$. Let $\Pi_k^*(\rho)$ and $X_{k-1}^*(\rho)$ be the optimal solutions of the subproblems for $k = 1$ to N. We can evaluate

$$X_{k-1}^* - \Pi_k^* = \varepsilon_k^* \tag{5.181}$$

If $\varepsilon_k = 0$ for all k, the global solution is obtained; if not, the interconnection constraint is not satisfied, and after having taken into account the results of the first level, action at the second level is necessary in order to ensure the satisfaction of the constraint equation (5.181). The essential transfer of information using the nonfeasible or goal coordination method is shown in Figure 5.14. The algorithm used in the second level to update the interconnection multiplier is as follows:

$$\rho_k^{t+1} = \rho_k^t + K \cdot (X_{k-1}^* - \Pi_k^*), \qquad K > 0 \tag{5.182}$$

where t is the iteration number.

The values X_{k-1}^* and Π_k^* are optimal solutions from the first level. Equation (5.182) is a gradient-type algorithm in which K determined the convergence rate.

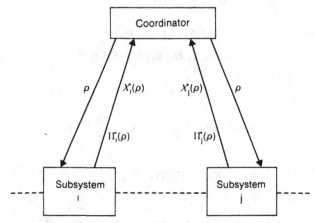

Figure 5.14. Information transfer of the nonfeasible method.

The feasible direction method that we employ for subproblem optimization can be outlined as follows:

Step 1. Let x_k be a feasible point (i.e., a point that satisfies the constraints).

Step 2. Find a direction S_k that is a feasible direction (i.e., a direction that has the property that there exists some $\lambda > 0$ such that

$$X_{k+1} = X_k + \lambda \cdot S_k \qquad (5.183)$$

will reduce the objective function and will be feasible for all $\lambda[0, \bar{\lambda}])$.

Step 3. Find a step size λ^* that minimizes the objective function (or find a point close to the minimum) along the direction S_k subject to the requirement that

$$X_{k+1} = X_k + \lambda^* \cdot S_k \qquad (5.184)$$

is feasible.

Step 4. Stop if X_{k+1} satisfies the convergence criteria; otherwise return to step 2.

The directions obtained by linear programming are tangent to the equality constraints, and any feasible points remain in the feasible region for a problem with linear equality constraint. For a problem that contains nonlinear equality constraints, however, there is no feasible direction that satisfies the nonlinear equality constraints. To illustrate, consider Figure

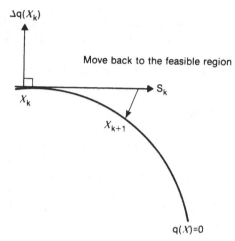

Figure 5.15. Illustration of the nonlinear equality constraint.

5.15, where a single equality constraint is shown. The feasible direction S_k is tangent to the constraint function $q(X_k)$ at point X_k. No matter how small a step is taken along S_k, the nonlinear equality constraint will be violated. For such problems, it is necessary to relax the requirements of feasibility by allowing the point to deviate slightly from the constraint and subsequently moving back to the feasible region.

To be specific, consider the following probem:

$$\min J(X) \tag{5.185}$$

subject to

$$g_i(X) \leq 0, \qquad \text{for } i = 1, \ldots, m \tag{5.186}$$

$$q_i(X) = 0, \qquad \text{for } i = 1, \ldots, n \tag{5.187}$$

Let X_k be a feasible point and I an active constraint set; i.e., $I = \{i: g_i(X_k) = 0\}$. Solve the following linear program:

$$\min \nabla J(X_k)^{\mathrm{T}} \cdot S_k \tag{5.188}$$

subject to

$$\nabla g_i(X_k)^{\mathrm{T}} \cdot S_k \leq 0, \qquad \text{for } iI \tag{5.189}$$

$$\nabla q_i(X_k)^{\mathrm{T}} \cdot S_k = 0, \qquad \text{for } i = 1, \ldots, n \tag{5.190}$$

$$-D < S_k < D \tag{5.191}$$

where $D = [1, 1, \ldots, 1]^T$. Equation (5.191) represents a normalizing condition that ensures a bounded solution. The resulting direction S_k is tangent to the equality constraints and to some of the binding nonlinear inequality constraints. A search along S_k is performed, and then a move back to the feasible region leads to X_{k+1} and the process is repeated.

5.4.2. Problem Formulation

In this section, we discuss the application of the previous technique to a simplified core mode. The assumed core contains 4×4 fuel rod assemblies with a single control rod located in the center. Each assembly has 24 nodes along the axis, and each node is assigned a xenon and an iodine concentration. Control variables include core flow and control rod position.

The objective or "cost" function is defined as

$$J = \int_{t_0}^{t_f} \{Q_p(t)[P^d(T) - P(t)]^2$$

$$+ [U(t) - U^d(T)]^T Q_u[U(t) - U^d(t)]\} \, dt$$

$$+ [X(t_f) - X(t_0)]^T Q_x[X(t_f) - X(t_0)] \qquad (5.192)$$

$P(t)$, $P^d(t)$ = generated and desired total thermal power, respectively
$U(t)$, $U^d(t)$ = derived and desired control input, respectively; they are two-dimensional vectors that model control rod and core flow
$X(t_f)$, $X(t_0)$ = Xe-I concentration at final and initial times, respectively; they are $2m$-dimensional vectors that correspond to the m nodes of the reactor core
Q_p, Q_u, Q_x = weighting factor matrices for power, control input, and the Xe-I concentration penalty, respectively; Q_u is a 2×2 matrix, and Q_x is a $2m \times 2m$ matrix. The Qs are positive definite; $Q_p(t)$ and $Q_u(t)$ are chosen to be time-dependent to express their relative importance during different parts of a load cycle.

The object of the cost function is to achieve the desired thermal power with minimal control effort and to return the reactor to as near the initial Xe-I concentration as possible. In addition, equation (5.192) requires the final penalty term in order to avoid an unbounded solution. Generally speaking, interchanging the control rod sequence is accomplished during the lower power level. It is thus required to achieve the desired control rod pattern when power decreases. At the full-power level, a specified control rod pattern is maintained because of fuel burnup considerations.

The discrete form of the objective function is

$$J = \sum_{k=1}^{N} [Q_{pk}(P_k^d - P_k)^2 + (U_k - U_k^d)^{\mathrm{T}} Q_{uk}(U_k - U_k^d)]$$

$$+ (X_N - X_0)^{\mathrm{T}} Q_x (X_N - X_0) \qquad (5.193)$$

where k represents the kth time step and 0 means initial value.

With the nonfeasible method, the objective function of the kth subsystem is

$$J_k = Q_{pk}(P_k^d - P_k)^2 + (U_k - U_k^d)^{\mathrm{T}} Q_{uk}(U_k - U_k^d) + \rho_{k+1}^{\mathrm{T}} \cdot X_k - \rho_k^{\mathrm{T}} \cdot \Pi_k$$

$$(5.194)$$

where Π_k is the interconnection parameter representing the initial xenon or iodine concentration of the kth subsystem.

In equation (5.193), generated power is the key concern, and thus it is heavily weighted. The weighting of the control term can be changed to adjust the control effort. The Q_x is adjusted so that a bounded Π_N can be obtained.

The Xe-I dynamics are described by the equations

$$\frac{dN_x^l(t)}{dt} = -\lambda_x N_x^l(t) + \lambda_I N_I^l(T) - \sigma_a^l N_x^l(t)\Phi^l(t) + \gamma_x \Sigma_f^l \Phi^l(t) \quad (5.195)$$

and

$$\frac{dN_I^l(t)}{dt} = -\lambda_I N_I^l(t) + \gamma_I \Sigma_f^l \Phi^l(t) \qquad (5.196)$$

where

$N_x^l(t) =$ xenon concentration of node l at time t
$N_I^l(t) =$ iodine concentration of node l at time t
$\lambda_x, \lambda_I =$ decay constants of xenon and iodine, respectively
$\gamma_x, \gamma_I =$ fissions yields of xenon and iodine, respectively
$\Phi^l(t) =$ neutron flux of node l at time t
$\Sigma_f^l =$ fission cross section of node l
$\sigma_a^l =$ xenon microscopic absorption cross section of node l

The following assumptions are used when equations (5.195) and (5.196) are discretized.

1. The total power change is approximated as shown in Figure 5.16, where the step power shape is used in the integration.
2. The nodal source distribution is assumed to be equal to the value of the previous time step.
3. The absolute value of the local power is scaled in proportion to the change in total power.

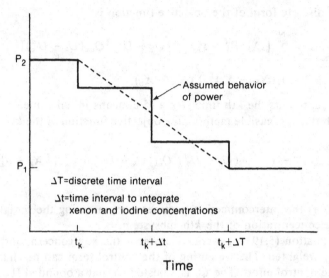

Figure 5.16. Illustration of assumed power behavior during Xe-I integration.

4. The xenon absorption cross section is assumed to be equal to the value at time $t_k + \Delta T$.

The xenon and iodine concentrations are updated step by step by using identical equations as follows:

$$N_x(t_k + \Delta t) = \frac{(\gamma_I + \gamma_x)\Sigma_f \Phi(t_k)}{\lambda_x + \sigma_a \Phi(t_k)}$$

$$+ \exp\{-[\lambda x + \sigma_a \Phi(t_k)]\Delta t\}\left[N_x(t_k) - \frac{\Sigma_f \Phi(t_k)(\gamma_I + \gamma_x)}{\lambda_x + \sigma_a \Phi(t_k)} \right.$$

$$\left. - \frac{\lambda_I N_I(t_k) - \gamma_I \Sigma_f \Phi(t_k)}{\lambda_x - \lambda_I + \sigma_a \Phi(t_k)} \right]$$

$$+ \exp(-\lambda_I \Delta t) \frac{\lambda_I N_I(t_k) - \gamma_I \Sigma_f \Phi(t_k)}{\lambda_x - \lambda_I + \sigma_a \Phi(t_k)} \qquad (5.197)$$

and

$$N_I(t_k + \Delta_t) = N_I(t_k) \exp(-\lambda_I \Delta t) + \frac{\gamma_I}{\lambda_I}\Sigma_f \Phi(t_k)[1 - \exp(-\lambda_I \Delta t)] \qquad (5.198)$$

where superscript l has been omitted for clarity and $\Phi(t_k)$ is defined as

$$\Phi = \frac{P(MW) \times 3.16 \times 10^{16} \text{ fission/MW} \times \text{constant}}{\text{Number of nodes} \times \text{volume of node} \times \Sigma_f} \times \psi \qquad (5.199)$$

where ψ is the nodal source. The value of Φ in equations (5.197) and (5.198) used the value of ψ at time t_k and total power P at time $t_k + \Delta t$. The constant is used to convert nodal source to nodal power.

The interconnection variables Π are defined as

$$\Pi_1 = N_x(t_k) \qquad (5.200)$$

and

$$\Pi_2 = N_1(t_k) \qquad (5.201)$$

and the state variables X as

$$X_1 = N_x(t_k + \Delta t) \qquad (5.202)$$

and

$$X_2 = N_1(t_k + \Delta t) \qquad (5.203)$$

Equations (5.197) and (5.198) are rewritten as

$$f_1 = \frac{(\gamma_1 + \gamma_x)\Sigma_f\Phi}{\lambda_x + \sigma_a\phi} + \exp[-(\lambda_x + \sigma_a\Phi)\Delta t]$$
$$\times \left[\Pi_1 \frac{(\gamma_1 + \gamma_x)\Sigma_f\Phi}{\lambda_x + \sigma_a\Phi} - \frac{\lambda_1\Pi_2 - \gamma_1\Sigma_f\Phi}{\lambda_x - \lambda_1 + \sigma_a\Phi}\right]$$
$$+ \exp(-\lambda_1\Delta t)\frac{\lambda_1\pi_2 - \gamma_1\Sigma_f\phi}{\lambda_x - \lambda_1 + \sigma_a\phi} - X_1$$
$$= 0 \qquad (5.204)$$

and

$$f_2 = \Pi_2 \exp(-\lambda_1\Delta t) + \frac{\gamma_1}{\lambda_1}\Sigma_f\Phi[1 - \exp(-\lambda_1\Delta_t)] - X_2 = 0 \quad (5.205)$$

Equations (5.204) and (5.205) are treated as equality constraints in the optimization problem. Therefore, F_k of equation (5.173) consists of $f_1^1, f_2^1, f_1^2, f_2^2, \ldots, f_1^l, f_2^l, \ldots, f_1^m, f_2^m)^T$, X_k of equation (5.173) consists of $(X_1^1, X_2^1, \ldots, X_1^m, X_2^m)^T$, and Π_k of equation (5.173) consists of $(\Pi_1^1, \Pi_2^1, \ldots, \Pi_1^l, \Pi_2^l, \ldots, \Pi_1^m, \Pi_2^m)^T$.

The nodal source is obtained from the nodal model NODE-B/THERM-B, the details of which are described in Reference 5.26.

The inequality constraints considered in this analysis are expressed in terms of the average and maximum planar heat generation rates, the linear heat generation rate (LHGR), the preconditioned fuel envelope, and the load-line limits from the operational power-flow map.

The coordination scheme is given by a gradient-type algorithm used to update the Lagrange multiplier; i.e.,

$$\rho_{ik}^{t+1} = \rho_{ik}^{t}\left[1 + K\left(1 - \frac{\Pi_{ik}^{*}}{X_{ik-1}^{*}}\right)\right], \qquad K > 0$$

where t = iteration number
         i = ith component of the vector
         k = kth subsystem
Π_{ik}^{*}, X_{ik-1}^{*} = optimal solutions from the first level
         K = predetermined constant

The above relation is used to update ρ_k for all the nodes. The terminology of subsystem and level is illustrated in Figure 5.13.

Partial derivatives are required when the linear programming problem described in Section 5.4.2 is to be solved. The scaled variables of the problem include the interconnections Π, the Xe-I concentrations X, the nodal source ψ, the total power P, core flow W, and control rod position R. Partial derivatives of the objective function, equality constraints, and inequality constraints are formed with respect to these variables.

5.4.3. Solution Algorithm

The final computer program contains portion of the neutronic/thermal code NODE-B/THERM-B and a linear programming code MINOS. The calculational sequence shown in Figure 5.17 is as follows:

1. Read core data such as albedo, B constants, initial conditions of local power, Xe-I concentration, etc.

2. Calculate the initial feasible point of each subsystem. For a new starting case, the initial feasible point is based on a guessed power level and control rod position for which a flow search calculation is performed. For an intermediate iterative case, the initial feasible point is based on an improved core flow and control rod position for which a power search calculation is performed. For both cases, the initial local power and Xe-I concentrations of each subsystem are the result of the previous subsystem calculations. NODE-B has the capability to search for the core flow if the power and control rod position are defined. This process is called the *flow search calculation*. If, on the other hand, the core flow and control rod position are defined and the power is sought, the process is called the *power search*.

3. Calculate the partial derivatives of equality and inequality constraints.

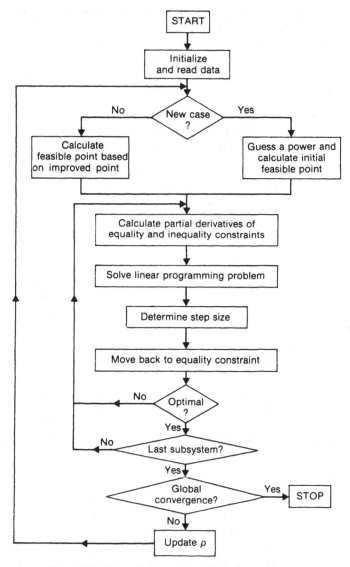

Figure 5.17. Flow diagram of the calculational sequence.

4. Solve the linear programming problem by calling MINOS.

5. Determine the step size along the feasible direction.

6. Set Π, core, flow, and control rod position equal to the improved value and call NODE-B to calculate a new feasible point that satisfies all the constraints.

7. Check to determine if the solution is optimal by showing that there is no further reduction in the optimal function. If not, repeat the iterative procedure by going back and recomputing the partial derivatives and other calculations. If the solution is optimal, optimize the next subsystem beginning with the calculation of the partial derivatives.

8. Check to see if global convergence is achieved. The criterion by which global convergence is determined is that each component of $X^*_{k-1} - \Pi^*_k$ is smaller than a small preset value for all the subsystems.

9. Update the Lagrange multiplier ρ if convergence is not achieved. Repeat the procedure, beginning with the calculation of the feasible point of each subsystem.

5.4.4. Results

The case presented here is based on a 4 × 4 assembly core with a control rod located at the center. By proper adjustment of the albedo of the boundary assemblies, the assumed reactor core behaves as if it were composed of an infinite number of identical units. The NODE-B approach takes advantage of the ability to run quarter-core calculations instead of full-core calculations if the core configuration has mirror symmetry. The actual number of nodes used in the problem treated here is 96, with 24 nodes in the axial direction. Specific power plant data used in the problem (such as B constants) are from Oyster Creek's cycle 9 in order to simulate practical operations.

The reactor is initially at full power and full-core flow rate with the control rod at node 6. The daily load cycle assumed is as follows. Power is reduced to 50% in 1 h, remains on that level for 6 h, and is returned to full power for the final 16 h. The 24-h time interval is divided into 11 time steps: 1, 2, 2, 2, 1, 2, 2, 3, 3, 3, and 3 h, respectively. The specified control rod position during the low-power period is at node 7.

The objective functional is formulated so that there are large penalties on deviation from the desired total power, as well as on deviation of the control rod from its desired position. The weighting of the penalty term is chosen to possess the same order as that of the Lagrange multiplier multiplied by the equilibrium xenon or iodine concentrations. The initial Lagrange multipliers are selected so that the sum of their product over the equilibrium xenon or iodine concentration is comparable to the power deviation term.

Figure 5.18 shows the optimal control strategy and also the average xenon concentration. The control rod is moved to the desired position, and core flow is adjusted to achieve the desired load demand. The variations of core flow are performed to balance xenon reactivity. The capacity loss

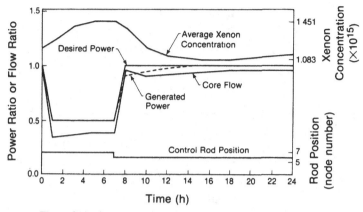

Figure 5.18. Control strategy and average xenon concentration.

of the reactor is due to xenon redistribution when the power is raised. Table 5.2 shows the LHGR envelope of hot points, with the value in parentheses showing the ramp rate value of the hottest point. The ramp rate value imposes a limitation on power increase. The deviation of power from the desired obtained with this strategy is about 0.5%, an acceptable result in view of the accuracy provided by the nodal method. Figures 5.19 through 5.22 show some interesting iterative details of subsystems 3, 5, 7, and 10, respectively. The selected parameters are the average xenon and iodine concentrations of the initial and final states of the subsystems, and core

Table 5.2. The LHGR Envelope and Envelope Ramp Rate of Hot Points of the Hottest Assembly $(2, 2)$

$(1, 1)$	$(1, 2)$
$(2, 1)$	$(2, 2)$

	Coordinate				
Subsystem number	$(6, 2, 2)$	$(7, 2, 2)$	$(8, 2, 2)$	$(9, 2, 2)$	$(10, 2, 2)$
5			10.277 (0.277)	10.229	
6		10.618 (0.309)	10.737	10.491	10.052
7	10.616 (0.308)	11.123	10.796	10.306	

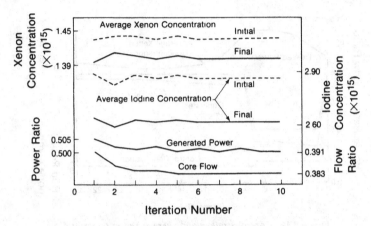

Figure 5.19. Iterative detail of subsystem 3.

flow and power. Figure 5.19 shows that the variations of all the parameters are relatively small and that they converge at the sixth iteration. The variations of parameters shown in Figure 5.20 are relatively large and also converge at the sixth iteration. Since the LHGR envelope is close to the allowable value at subsystem 7, the improvement achieved in each iteration as shown in Figure 5.21 is more limited. Thus, the relative variations of the parameters determined the speed of calculation. The improvements of subsystem 10 (Figure 5.22) continued after the sixth iteration. Since the difference of Π_k and X_{k-1} is within 3% in this case, however, the final global convergence of all the parameters is not affected.

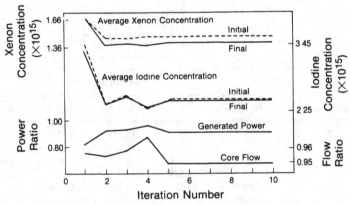

Figure 5.20. Iterative detail of subsystem 5.

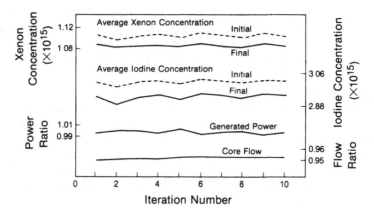

Figure 5.21. Iterative detail of subsystem 7.

5.4.5. Concluding Remarks

The highly heterogeneous characteristics of a BWR, which are accentuated by the discrete movement of control rods and the strong effect of the control rods on the power distribution, suggest the importance of detailed spatial information, requiring a rather detailed neutronic/thermal reactor model to properly describe the reactor behavior. In this study we have used the NODE-B/THERM-B model from the Electric Power Research Institute, which has performed well in monitoring the reactor core power shape. The dimensions of the problem, however, are still extremely large, and the multilevel method appears to be a practical way to overcoming this difficulty.

Generally speaking, the rate of convergence of the nonfeasible method is relatively slow, but it has the advantage of placing no restrictions on the

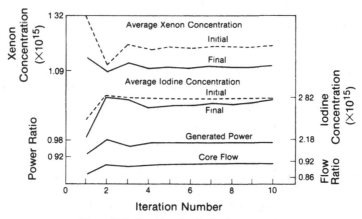

Figure 5.22. Iterative detail of subsystem 10.

characteristics of the problem. Since the Lagrange multiplier is formulated in the objective function, its effect also depends on the value of the other terms. The initial Lagrange multiplier does not influence the final optimal control strategy, but it influences the convergent path. The K-value of the gradient-type coordination algorithm has no effect on the final results but does affect the iteration path.

The feasible direction method satisfactorily performs the optimization of subproblems. It is possible that the solution may move toward a local minimum instead of a globally optimal solution due to the large number of variables and the nonlinearity of the problem. Since the final desired results, however, are control position, core flow, and power, the difficulty can be overcome by heavily weighting the power deviation and control rod position deviation terms.

The method of optimal control suggested in this section depends on a detailed spatially dependent feedback model and calculational procedure. For this reason it is more suitable for design studies rather than for on-line control. It is possible that a simpler coarse-mesh dynamics model could be developed in conjunction with the hierarchical time decomposition that would be better suited to on-line application. A balance would have to be struck, however, between a physically simpler model and the incorporation of realistic inequality constraints based on materials and safety limits.

Appendix A

Equation (5.93), repeated here as equation (A.1),

$$Lu + \lambda u = g(x, u), \qquad \chi \in D$$
$$u = 0, \qquad \chi \in \partial D \tag{A.1}$$

is a nonlinear eigenvalue problem. We begin the analysis of this class of problems with the following.

Definition. The *branch point* (*bifurcation point*) of equation (5.93) is that value of λ, say λ_0, for which there is a solution, $u_0(\chi)$, and such that λ_0 is the fundamental eigenvalue of

$$L\psi_0 + \lambda_0\psi_0 = g_u[\chi, u_0(\chi)]\psi_0, \qquad \chi \in D$$
$$\psi_0 = 0, \qquad \chi \in \partial D \tag{A.2}$$

and for which $u_0(\chi; \lambda_0)$ may split off into multiple solutions.

We note that $u_0(\chi) = 0$ is a solution of equation (A.1) for any λ, say λ_0, and further observe that $g_u(\chi, 0) = 0$ in equation (A.2). Therefore λ_0

and $\psi_0(\chi)$ can be found from equation (A.2) independently of equation (A.1). Hence, by definition, λ_0 is a branch point of equation (A.1) with corresponding $u_0(\chi) = 0$.

In order to show that equation (A.1) with $\hat{B} > 0$ has a unique positive solution, bifurcated from λ_0 and $u_0(\chi) = 0$, on the continuous spectrum $\lambda > \lambda_0$, we need a theorem and some observations on the nonlinear term $g(\chi, u)$.

We use the existence theorem first proved by Amann and reproduced here in substance for solutions of nonlinear boundary value problems of the type

$$Lu + f(\chi, u) = 0, \qquad x \in D,$$
$$u = 0, \qquad x \in \partial D$$
$$\text{(A.3)}$$

where L is a second-order uniformly elliptic operation

$$L = \sum_{i,j=1}^{n} a_{ij}(\chi) \frac{\partial^2}{\partial \chi_i \, \partial x_j} - a_0(\chi), \qquad \chi = (\chi_1, \chi_2, \ldots, \chi_n)$$

The coefficients $a_{ij}(\chi)$ of L as well as $a_0(\chi)$ are assumed to be Hölder continuous with exponent α and $a_0(\chi) \geq 0$. We assume that ∂D belongs to the class of $C^{2+\alpha}$.

Theorem. Let there exist two smooth functions $u_0(\chi) \geq v_0(\chi)$ in C^{2+a} such that

$$Lu_0 + f(\chi, u_0) \leq 0, \qquad x \in D$$
$$u_0 \geq 0, \qquad \chi \in \partial D$$
$$\text{(A.4)}$$

and

$$Lv_0 + f(\chi, v_0) \geq 0, \qquad \chi \in D,$$
$$v_0 \leq 0, \qquad \chi \in \partial D$$
$$\text{(A.5)}$$

Further assume that $f(\chi, u)$ is Hölder continuous in χ and in u on $\min v_0 \leq u \leq \max u_0$. Then there exists at least one solution u of equation (A.3) satisfying the inequality

$$v_0 \leq u \leq u_0 \qquad \text{(A.6)}$$

The proof is based on constructing two monotone approximation sequences: an increasing one starting from v_0, called a *lower solution*, and converging to $\underline{u}$, and a decreasing one starting from u_0, called an *upper solution*, and converging to $\bar{u}$, where $\underline{u}$ and $\bar{u}$ are minimal and maximal solutions of (A.3), respectively.

From the assumption of $f(\chi, u)$, there exists a $\Omega(\chi)$ in C^α satisfying

$$\Omega(\chi) + a_0(\chi) \geq 0 \tag{A.7}$$

and

$$f(\chi, \psi) - f(\chi, \Phi) \geq -\Omega(\chi)(\psi - \Phi) \tag{A.8}$$

for min $v_0 \leq \Phi \leq \psi \leq$ max u_0.

We define a mapping T,

$$v = Tu \tag{A.9}$$

such that

$$(L - \Omega)v = -[f(\chi, u) + \Omega u], \qquad x \in D \tag{A.10}$$

$$v = 0, \qquad \chi \in \partial D \tag{A.11}$$

Then we can show that T is a monotone operator ($u \leq v$ implies $Tu < Tv$) provided min $v_0 \leq u$, $v \leq$ max u_0. Suppose $u \leq v$. Then

$$(L - \Omega)Tu = -[f(\chi, u) + \Omega u], \qquad \chi \in D \tag{A.12}$$

$$Tu = 0, \qquad \chi \in \partial D \tag{A.13}$$

$$(L - \Omega)Tv = -[f(\chi, v) + \Omega v], \qquad \chi \in D \tag{A.14}$$

$$Tv = 0, \qquad \chi \in \partial D \tag{A.15}$$

Therefore,

$$(L - \Omega)(Tv - Tu) = -[f(\chi, v) - f(\chi, u) + \Omega(v - u)], \qquad \chi \in D \tag{A.16}$$

$$Tv - Tu = 0, \qquad \chi \in \partial D \tag{A.17}$$

Using equation (A.8), we get

$$(L - \Omega)(Tv - Tu) \leq 0, \qquad \chi \in D \tag{A.18}$$

$$Tv - Tu = 0, \qquad \chi \in \partial D \tag{A.19}$$

By the strong maximum principle for elliptic equations (Ref. 5.28), we get

$$Tv > Tu, \qquad \chi \in D \tag{A.20}$$

unless $Tv = Tu$, in which case the right side of equation (A.16) is identically zero, but this happens only if $u \equiv v$. Therefore, T is a monotone operator.

Now we define a sequence

$$u_n = TU_{n-1}, \qquad n = 1, 2, \ldots \tag{A.21}$$

For $n = 1$, we have

$$(L - \Omega)u_1 = -[f(\chi, u_0) + \Omega u_0], \qquad \chi \in D, \tag{A.22}$$

$$u_1 = 0, \qquad \chi \in \partial D \tag{A.23}$$

so

$$(L - \Omega)(u_1 - u_0) = -[Lu_0 + f(\chi, u_0)], \qquad \chi \in D \qquad (A.24)$$

Using equations (A.4) and (A.23), we get

$$(L - \Omega)(u_1 - u_0) \geq 0, \qquad \chi \in D, \qquad (A.25)$$

$$u_1 - u_0 \leq 0, \qquad \chi \in \partial D \qquad (A.26)$$

Again, by the strong maximum principle,

$$u_1 < u_0, \qquad \chi \in D \text{ [assume } Lu_0 + f(\chi, u_0) \neq 0] \qquad (A.27)$$

Since $u_1 < u_0$ and T is monotone,

$$u_2 = Tu_1, \qquad u_1 = Tu_0$$

Hence $u_2 < u_1$. Thus the sequence defined by equation (A.21) is monotone decreasing, and, analogously, $v_n = Tv_{n-1}$ defines a monotone increasing sequence

$$v_0 < v_1 < v_2 < \cdots < u_2 < u_1 < u_0 \qquad (A.28)$$

Now $v_n < u_n$ is shown by induction:

1. From the assumption, $v_0 < u_0$.
2. Suppose $v_{n-1} < u_{n-1}$.
3. Then $Tv_{n-1} < Tu_{n-1}$ since T is monotone.

Hence $v_n < u_n$.

Since the sequences $\{u_k\}$ and $\{v_k\}$ are monotone, the pointwise limits

$$\bar{u}(\chi) = \lim_{k \to \infty} u_k(\chi) \qquad \text{and} \qquad \underline{u}(\chi) = \lim_{k \to \infty} v_k(\chi)$$

both exist.

Returning to equation (A.1), we identify $f(\chi, u) = \lambda u - g(\chi, u)$ and observe the following properties of $g(\chi, u)$ with $\hat{B} > 0$;

1. $g(\cdot, 0) = 0$.
2. $g_u(\cdot,) = 0$.
3. $g(\cdot, u) > 0$ for $u > 0$.

4. $\displaystyle \lim_{u \to \infty} \frac{g(\cdot, u)}{u} = +\infty$.

5. $\dfrac{g(\cdot, u)}{u}$ strictly increases for $u > 0$.

If we choose $v_0 = \varepsilon\psi_0$, where ε is a positive constant and ψ_0 is the positive fundamental eigenfunction corresponding to λ_0, then

$$Lv_0 + f(\chi, v_0) = L\varepsilon\psi_0 + \lambda\varepsilon\psi_0 - g(\chi, \varepsilon\psi_0)$$

$$= (\lambda - \lambda_0)\varepsilon\psi_0 - g(\chi, \varepsilon\psi_0), \qquad \chi \in D$$

$$\varepsilon\psi_0 = 0, \qquad \chi \in \partial D \qquad (A.29)$$

Due to properties 1, 2, and 3, for any $\lambda > \lambda_0$, we can choose ε sufficiently small so that

$$(\lambda - \lambda_0)\varepsilon\psi_0 - g(\chi, \psi_0) > 0, \qquad \chi \in D \tag{A.30}$$

Therefore, we can find v_0, a lower solution that satisfies equation (A.5).

Next, if we choose $u_0 = C$, where C is a positive constant, then

$$Lu_0 + f(\chi, u_0) = -a_0(\chi)C + \lambda C - g(\chi, D), \qquad \chi \in D,$$
$$C > 0, \qquad \chi \in \partial D \tag{A.31}$$

Due to property 4 for any given λ, we can choose C large enough so that

$$-a_0(\chi)C + \lambda C - g(\chi, C) < 0, \qquad \chi \in D \tag{A.32}$$

Therefore, we can find u_0, an upper solution that satisfies equation (A.4).

Depending on the choice of $\Omega(\chi)$, we have a variety of monotone iteration schemes as long as the $\Omega(\chi)$'s are chosen to satisfy equations (A.7) and (A.8). Since $f(\chi, u) = \lambda u - g(\chi, u)$ in our problem, equation (A.8) is equivalent to, for $\min v_0 \leq \Phi \leq \Psi \leq \max u_0$,

$$\lambda\psi - g(\chi, \psi) - \lambda\Phi + g(\chi, \Phi) \geq -\Omega(\chi)(\Psi - \Phi)$$

or

$$\lambda - \frac{g(\chi, \Psi) - g(\chi, \Phi)}{\Psi - \Phi} \geq -\Omega(\chi) \tag{A.33}$$

Hence, a possible choice of $\Omega(\chi)$ is

$$\Omega(\chi) = -\lambda + g_u(\chi, \cdot) + N\lambda_0, \qquad N = 0, 1, 2, \ldots \tag{A.34}$$

To ensure that equations (A.33) and (A.7) are satisfied, $N\lambda_0$ is added. If we take $g_u(\chi, \cdot) = g_u[\chi, u_{n-1}(\chi)]$, the iteration scheme becomes essentially the process of quasilinearization.

Hence, by the theorem, equation (A.1) has at least one positive solution for $\lambda > \lambda_0$.

Since $\underline{u}$ and $\bar{u}$ are minimal and maximal solutions of equation (A.1) generated by starting from $\varepsilon\psi_0$ and C, respectively, we have

$$L\underline{u} + \lambda\underline{u} = g(\chi, \underline{u}), \qquad \chi \in D$$
$$\underline{u} = 0, \qquad \chi \in \partial D \tag{A.35}$$

$$L\bar{u} + \lambda\bar{u} = g(\chi, \bar{u}), \qquad \chi \in D$$
$$\bar{u} = 0, \qquad \chi \in \partial D \tag{A.36}$$

Multiplying equation (A.35) by $\bar{u}$ and equation (A.36) by $\underline{u}$ and subtracting, we have

$$\int_D \underline{u}\bar{u}\left[\frac{g(\chi, \underline{u})}{\underline{u}} - \frac{g(x, \bar{u})}{\bar{u}}\right] d\chi = 0 \tag{A.37}$$

for a self-adjoint L.

By item 5 and since $\bar{u} \geq \underline{u} > 0$, we conclude that

$$\underline{u} = \bar{u} \tag{A.38}$$

that is, the positive solution is unique.

Hence, it is enough to generate only one sequence. In this case, it is obvious that we will choose the sequence starting from $v_0 = \varepsilon\psi_0$, since ψ_0 already satisfies the boundary condition.

Appendix B

The system of equations (5.136) and (5.137) is a linear two-point boundary value problem of the form

$$Lx(t) \triangleq \dot{x} - K(t)x = f_0(t) \tag{B.1}$$

$$Ax(t_0) = \alpha \tag{B.2}$$

$$Bx(t_f) = \beta \tag{B.3}$$

where A is an $(n - q) \times n$ matrix of rank $n - q$, B is a $q \times n$ matrix of rank q, α is a vector with $n - q$ components, and β is a vector with q components.

For the above problem, the computer code SUPORT utilizes

1. Superposition coupled with orthonormalization
2. State-of-the-art of initial value problem solvers
3. Reduction of the number of equations to be solved by suitably choosing initial values at t_0

Superposition is perhaps the simplest of all methods for solving linear boundary value problems. The method proceeds by producing linearly independent solutions of the homogeneous equation and a particular solution of the inhomogeneous equation. The solution of the original problem is then produced by forming an appropriate linear combination that satisfies the given boundary conditions. First, solve $q + 1$ (not $n + 1$) systems of initial value problems:

$$L\chi_0(t) = f_0(t), \qquad A\chi_0(t_0) = \alpha \tag{B.4}$$

$$L\chi_\nu(t) = 0, \quad A\chi_\nu(t_0) = 0, \qquad \nu = 1, 2, \ldots, q \tag{B.5}$$

Seek as the solution a linear combination of the form

$$\chi(t) = \chi_0(t) + \sum_{v+1}^{a} c_v \chi_v(t)$$

or

$$\chi(t) = \chi_0(t) + X(t)c \tag{B.6}$$

Notice that, in equations (B.4) and (B.5), the initial conditions for $x(t)$ have been chosen to satisfy (B.2). Now c is determined by (B.3):

$$B\chi(t_f) = B\chi_0(t_f) + BX(t_f)c = \beta$$

or

$$B\chi(t_f)BX(t_f)c = \beta - B\chi_0(t_f) \tag{B.7}$$

Determination of c is unique if and only if $BX(t_f)$ is nonsingular (i.e., if the original problem has a unique solution). The original two-point boundary problem has been transformed into systems of initial value problems and a matrix algebraic equation. Various sophisticated integrators for the initial value problems can be used here.

In order for the method to yield accurate results, it is important that $\chi_0(t)$ and the columns of $X(t)$ in equation (B.6) be linearly independent for all t. Practically, since computers use finite word length, the solutions may lose numerical independence. To overcome this, superposition is combined with an orthonormalization procedure. Independence of the solutions over the entire interval is preserved by ensuring that they are nearly mutually orthogonal. Each time the solution vectors start to lose their numerical independence, the code reorthonormalizes the homogeneous solution vectors and turns the particular solution into the orthogonal complement to the new homogeneous solution vectors by a modified Gram–Schmidt procedure.

If t_m represents an orthonormalization point, a new orthonormal set of vectors is formed and becomes the initial conditions of the integration subinterval $[t_m, t_{m+1}]$:

$$X_{m-1}(t_m) = X_m(t_m)P_m \tag{B.8}$$

and

$$\chi_{0m}(t_m) = \chi_{0m-1}(t_m) - X_m(t_m)w_m \tag{B.9}$$

where Gram–Schmidt formulation P_m and w_m can be found in Reference 5.24.

The superposition becomes

$$\chi_m(t) = \chi_{0m}(t) + X_m(t)c_m, \qquad t \in [t_m, t_{m+1}] \tag{B.10}$$

and c_m is obtained from continuity of the solutions

$$\chi_{m-1}(m) = \chi_m(t_m) \tag{B.11}$$

References

5.1. YOON, M. H., and NO, C. H., "Direct Numerical Technique of Mathematical Programming for Optimal Control of Xenon Oscillation in Load Following Operation," *Nucl. Sci. Eng.* **90**, 203-220 (1985).

5.2. KARPPINEN, J., VESLUIS, R. M., and BLOMSNES, B., "Core Control Optimization for Scheduled Load Changes in Large Pressurized Water Reactors," *Nucl. Sci. Eng.* **71**, 1-17 (1979).

5.3. CHO, N. Z., and GROSSMAN, L. M., "Optimal Control for Xenon Spatial Oscillations in Load Follow of a Nuclear Reactor," *Nucl. Sci. Eng.* **83**, 136-148 (1983).

5.4. LIN, C., and GROSSMAN, L. M., "Optimal Control of a Boiling Water Reactor in Load-Following via Multilevel Methods," *Nucl. Sci. Eng.* **92**, 531-544 (1986).

5.5. OLSSON, G., "Power-Nuclear Plants, Commentator's Report," Proceedings of the IFAC 5th World Congress, Session 6, Paris (1972).

5.6. TZAFESTAS, S., and CHRYSOCHOIDES, N., "Nuclear Reactor Control Using Walsh Function Variational Synthesis," *Nucl. Sci. Eng.* **62**, 763 (1977).

5.7. LEWINS, J., and BABB, A. L., "Optimum Nuclear Reactor Control Theory," *Adv. Nucl. Sci. Tech.* **4**, 252 (1968).

5.8. WIBERG, D. M., "Optimal Feedback Control of Spatial Xenon Oscillations," *Trans. Am. Nucl. Soc.* **7**, 219 (1966).

5.9. WIBERG, D. M., "Optimal Feedback Control of Spatial Xenon Oscillations in Nuclear Reactors," *Nucl. Sci. Eng.* **27**, 600 (1967).

5.10. STACEY, W. M., "Optimal Control of Xenon-Power Spatial Transients," *Nucl. Sci. Eng.* **33**, 162 (1968).

5.11. STACEY, W. M., "Control of Xenon Spatial Oscillations," *Nucl. Sci. Eng.* **38**, 229 (1969).

5.12. STACEY, W. M., "Application of Variational Synthesis to the Optimal Control of Spatially Dependent Reactor Models," *Nucl. Sci. Eng.* **39**, 229 (1970).

5.13. CHAUDHURI, S. P., "Distributed Optimal Control in a Nuclear Reactor," *Int. J. Control* **16**(5), 927 (1972).

5.14. LAZAREVIC, B., OBRADOVIC, D., and CUK, N., "Modal Approach to the Optimal Control System Synthesis of a Nuclear Reactor," *Int. J. Control* **16**(5) 817 (1972).

5.15. PERDIKIS, G. C., and BAILEY, R. E., "A New Boron Control Law for Load Following PWR's," *Trans. Am. Nucl. Soc.* 427 (1976).

5.16. ASSATANI, K., "Near Optimal Control of Distributed Parameter Systems Via Singular Perturbation Theory," *J. Math. Anal. Appl.* **54**, 799 (1976).

5.17. LOVE, C. G., "A Nonlinear Dynamic Optimization Technique for Controlling Xenon Induced Oscillations in Large Nuclear Reactors," *IEEE Trans. Nucl. Sci.* NS18(1), 408 (1971).

5.18. EBERT, D. D., TERNEY, W. B., WILLIAMSON, E. A., JR., and GOMM, N. R., "The Development of Maneuvering Strategies Using Optimal Control Theory," *Nucl. Sci. Eng.* **69**, 398 (1979).

5.19. BERAHA, D., and KARPPINEN, J., "Power Distribution Control by Hierarchical Optimization Techniques," Proceedings of the International Topological Meeting for Advances in Mathematical Methods for the Solution of Nuclear Engineering Problems, Munich, FRG, **2**, 331 (1981).

5.20. CHO, N. Z., "Optimal Control Theory for Xenon Spatial Oscillation in Load Follow of a Nuclear Reactor," Ph.D. thesis, University of California, Berkeley, 1980.

5.21. SAGE, A. P., and WHITE, C. C., *Optimum Systems Control*, Prentice-Hall, Englewood Cliffs, New Jersey, 1977.

5.22. HETRICK, D. L., *Dynamics of Nuclear Reactors*, University of Chicago Press, Chicago, 1971.

5.23. SCOTT, M. R., and WATTS, H. A., "SUPORT—A Computer Code for Two Point Boundary-Value Problems via Orthonormalization," SAND 75-0198, Sandia National Laboratories, 1975.

5.24. KAPLAN, S., "Synthesis Methods of Reactor Analysis," in *Advances in Nuclear Science and Technology*, Vol. 3, edited by P. Greebler and E. J. Henley, Academic Press, New York, 1966.

5.25. KELLER, J. B., and ANTMAN, D., *Bifurcation Theory and Nonlinear Eigenvalue Problems*, W. A. Benjamin, New York, 1969.

5.26. "Power Shape Monitoring System," Vol. 2, Technical Description, NP 1660, Electric Power Research Institute, 1981.

5.27. BERAHA, D., and KARPPINEN, K., "Power Distribution Control by Hierarchical Optimization Techniques", Proceedings of the International Topological Meeting for Advances in Mathematical Methods for the Solution of Nuclear Engineering Problems, Munich, FRG, **2**, 331 (1981).

5.28. JAMSHIDI, M., "Large Scale Systems," *Modelling and Control*, North-Holland, New York, 1983.

6

Application of the Minimum Norm Formulation to Problems in Control of Distributed Reactors

6.1. Introduction (Refs. 6.1–6.43)

In Chapter 5 we discussed applications of the minimum norm formulation to the search for the optimal control vector that minimizes a cost functional that penalizes the deviations from equilibrium along the trajectory and the required control effort. This chapter discusses applications of the minimum norm formulation to the search for the optimal control vector that minimizes the same cost functional but subject to satisfying one of the following conditions:

(1) At the end of a given time interval, the total power output from the reactor core should be equal to a specified value. This constraint is relevant to the problem of adjusting the power level while minimizing the control effort and the distortion of the flux distribution.

(2) The total power from the reactor core should match a specific load trajectory along a given time interval. This constraint is relevant to the problem of controlling a nuclear reactor core during load-following operations.

(3) At the end of the specified time interval, the state of the system should match a given distribution. Although this constraint is also relevant to the problem of adjusting the power output from the core, it is more restrictive than case (1).

6.2. The Nuclear Reactor Model

Although an accurate description of the power generated in a nuclear reactor core would have to take into account both the continuous energy spectra of the neutrons in the core and the energy-dependent fission cross section of the fuel, adequate results are often obtained through the simplifying assumption that the neutron population can be grouped into monoenergetic groups whose dynamics are described by multigroup diffusion theory.

The total power produced in a nuclear reactor core is given by

$$p(t) = \sum_{i=1}^{N_g} \int_V e_{f_i} \Sigma_{f_i}(r) \phi_i(r, t) \, dr \tag{6.1}$$

where the integral is over the volume of the reactor core, e_{f_i} is the energy released per fission involving a neutron in the ith energy flux ϕ_i, Σ_{f_i} is the fission macroscopic cross section for neutrons in the ith energy interval, and N_g denotes the number of energy intervals that cover the neutron energy range.

As in Chapter 4, we assume that near the equilibrium condition the state $\psi(r, t)$ of the nuclear reactor core is given by

$$\psi(r, t) = G(r, t; r', t_0) Z_0(r') + \int_{t_0}^{t} G(r, t; r', \tau) B(r') U(\tau) \, d\tau \tag{6.2}$$

where all the variables are defined as before.

In terms of the state $\psi(r, t)$, the total power $p(t)$ can be modeled by the expression

$$p(t) = \int_V \hat{H}(r) \psi(r, t) \, dr \tag{6.3}$$

where $\hat{H}(r)$ is a space-dependent row matrix with the appropriate dimension.

In what follows it is assumed that the state $\psi(r, t)$ and the control $U(t)$ belong to the Hilbert spaces H_2 and H_1, respectively. H_1 and H_2 are endowed with the inner products

$$\langle \psi, Z \rangle_{H_2} = \int_{t_0}^{t_1} \langle \psi(t), Z(t) \rangle_H \, dt \tag{6.4}$$

and

$$\langle U, W \rangle_{H_1} = \int_{t_0}^{t_1} k_0 \langle U(t) m W(t) \rangle_E \, dt \tag{6.5}$$

where k_0 is a positive real number. The inner products in H and E are defined by

$$\langle \psi(t), Z(t) \rangle_H = \int_V Z^T(r, t) Q(r) \psi(r, t) \, dr \tag{6.6}$$

and

$$\langle U(t), W(t) \rangle_E = W^T(r) R U(t) \tag{6.7}$$

in which R is an M-dimensional, positive-definite matrix, and $Q(r)$ is a space-dependent, N-dimensional, positive-definite matrix.

6.3. Optimal Control of the State Distribution with Power-Level Adjustment

If P_0 is the total output power corresponding to the initial state $Z_0(r)$ at time t_0. The problem is to find the control function $U(t)$ that changes the total output power from P_0 to P_1 in a given time interval $[t_0, t_1]$ and minimizes the cost functional

$$J(U) = \int_{t_0}^{t_1} \int_V [\psi(r, t) - Z(r, t)]^T Q(r) [\psi(r, t) - Z(r, t)] \, dr \, dt$$

$$+ k_0 \int_{t_0}^{t_1} U^T(t) R U(t) \, dt \tag{6.8}$$

where $Z(r, t)$ is the desired state distribution.

Although no explicit constraint is imposed on the total power output

$$p(t) = \int_V \hat{H}(r) G(r, t; r', t_0) Z_0(r') \, dr$$

$$+ \int_V \int_{t_0}^{t_1} \hat{H}(r) G(r, t; r', \tau) B(r') U(\tau) \, d\tau \, dr \tag{6.9}$$

along the given time interval, implicitly, the performance index (6.8) does penalize the output power variations along the trajectory.

The present problem includes on an equal footing the problem of controlling flux oscillations, minor power adjustments, and load-following operations. Defining the problem into any of these cases depends on both the time-scale and the reactor model considered.

To formulate the problem in the context of functional analysis, we introduce the following definitions: F denotes the transformation from H_1 into H_2 and is defined by

$$F(r, t; \tau) U(\tau) = \int_{t_0}^t G(r, t; r', \tau) B(r') U(\tau) \, d\tau \tag{6.10}$$

T denotes the transformation

$$T(t; \tau)U(\tau) = \int_{t_0}^{t} \int_{V} \hat{H}(r)G(r, t; \tau)B(r')U(\tau)\, dr\, d\tau \qquad (6.11)$$

Also, $y(r, t)$ and $\xi(t)$ are functions defined by

$$y(r, t) = Z(r, t) - G(r, t; r', t_0)Z_0(r') \qquad (6.12)$$

and

$$\xi(t) = p(t) - \int_{V} \hat{H}(r)G(r, t; r', t_0)Z_0(r')\, dr \qquad (6.13)$$

At time t_1 the function $\xi(t_1)$ is completely specified. The transformation T and the function ξ when evaluated at time t_1 will be denoted by T_1 and ξ_1. Finally, H_3 represents the one-dimensional real Hilbert space endowed with inner product

$$\langle \alpha, \beta \rangle_{H_3} = \alpha\beta \qquad (6.14)$$

With these definitions the optimal control problem can be formulated as follows: find the control U in H_1 that minimzes

$$J(U) = \|FU - y\|_{H_2} + \|U\|_{H_1} \qquad (6.15)$$

and satisfies

$$\xi_1 = T_1 U \qquad (6.16)$$

where T_1 is a linear and bounded transformation from H_1 onto H_3.

This formulation is Porter's abstract minimum norm problem.

6.3.1. Necessary and Sufficient Conditions of Optimality

By invoking Porter's result (Ref. 2.8), we find that the control U that minimzes the performance index (6.15) and satisfies (6.16) is given by

$$U = [I + F^*F]^{-1}(T_1^+\eta + F^*y) \qquad (6.17)$$

where η is the unique element of H_3 satisfying

$$T_1 U = \xi_1 \qquad (6.18)$$

The adjoint F^* transforms H_2 into H_1. The pseudoinverse T_1^+ associated with T_1 is given by

$$T_1^+ = T_1^*[T_1 T_1^*]^{-1} \qquad (6.19)$$

and the adjoint T_1^* is a transformation that maps H_3 into H_1.

Since there is no straightforward approach to computing the inverse of $[I + F^*F]$, it is necessary to express conditions (6.17) and (6.18) in a more standard form, amenable to computation. To this end, condition (6.17) is rewritten as

$$U = -F^*FU + T_1^+\eta + F^*y \tag{6.19a}$$

Also, given the algebraic property of the pseudoinverse

$$T_1T_1^+\eta = \eta \tag{6.20}$$

and using (6.18), we find from (6.19a) that

$$\xi = -T_1F^*FU + \eta + T_1F^*y \tag{6.21}$$

Solving for η from (6.21) and substituting it in (6.19a), we obtain the operator equation

$$U = [T_1^+T_1 - I]F^*FU + \Delta \tag{6.22}$$

in which the function Δ is an element of H_1, defined by

$$\Delta = T_1^+\xi - [T_1^+T_1 - I]F^*y \tag{6.23}$$

Note that (6.22) lends itself to application of successive approximation techniques. It is true that the pseudoinverse T_1^+ implicitly involves the inverse of the operator $T_1T_1^*$. However, since T_1 has finite-dimensional range, it follows that $T_1T_1^*$ is an operator with finite-dimensioal domain and range. More explicitly, $T_1T_1^*$ transforms the one-dimensional Hilbert space H_3 into itself. The inverse $[T_1T_1^*]^{-1}$, therefore, should pose no computational difficulties.

In order to fully characterize the optimality condition (6.22), we must still derive explicit expressions for F^* and T^+.

6.3.1.1. The Adjoint F^*

Following the same procedures as before, we can show that

$$F^*(\tau, t)Z(t) = k_0^{-1}\int_\tau^{t_1} B^*G^*(t; \tau)Z(t)\, dt \tag{6.24}$$

where for simplicity only the temporal variable is explicitly shown. The adjoint B^* transforms H into E according to

$$B^*Z = R^{-1}\int_V B^T(r)Q^T(r)Z(r)\, dr \tag{6.25}$$

and the adjoint G^* transforms H into itself.

Replacing $Z(t)$ in equation (6.24) by the defining relation (6.10), we obtain

$$F^*(\tau; t)F(t; \alpha)U(\alpha) = k_0^{-1}\int_\tau^{t_1} B^*G^*(t; \tau)\int_{t_0}^t G(t; \alpha)BU(\alpha)\, d\alpha\, dt \tag{6.26}$$

where again only the temporal dependency is explicitly shown.

After interchanging the order of integration in equation (6.26), we have

$$F^*(\tau; t)F(t; \alpha)U(\alpha) = k_0^{-1} \int_{t_0}^{t_1} K(\tau; \alpha)U(\alpha)\, d\alpha \qquad (6.27)$$

where

$$K(\tau; \alpha) = \begin{cases} \int_{\tau}^{t_1} B^*G^*(t; \tau)G(t; \alpha)B\, dt & \text{for } \alpha < \tau \qquad (6.28) \\ \int_{\alpha}^{t_1} B^*G^*(t; \tau)G(t; \alpha)B\, dt & \text{for } \alpha \geq \tau \qquad (6.29) \end{cases}$$

6.3.1.2. The Pseudoinverse T_1^+

The adjoint T_1^* is related to T_1 through the inner-product relation

$$\langle T_1 U, \xi \rangle_{H_3} = \langle U, T_1^* \xi \rangle_{H_1} \qquad (6.30)$$

Using equation (6.11), we expand the left side of (6.30) as follows:

$$\int_{t_0}^{t_1} \int_V \hat{H}(r)G(r, t_1; r', \tau)B(r')U(\tau)\, dr\, d\tau\, \xi \qquad (6.31)$$

which, in terms of the inner product in H, takes the form

$$\int_{t_0}^{t_1} \langle G(t_1; \tau)BU(\tau), Q^{-1}\hat{H}^T \xi \rangle_H\, d\tau \qquad (6.32)$$

From (6.32) it follows that

$$\langle T_1 U, \xi \rangle_{H_3} = \langle U, k_0^{-1}B^*G^*Q^{-1}\hat{H}^T \xi \rangle_{H_1} \qquad (6.33)$$

The adjoint T_1^* is therefore

$$T_1^*(\tau) = k_0^{-1}B^*G^*(t_1; \tau)Q^{-1}\hat{H}^T \qquad (6.34)$$

Substituting for B^* from equation (6.25) into equation (6.34), we get

$$T_1^*(\tau) = k_0^{-1}R^{-1} \int_V B^T(r)Q^T(r)G^*(r, t_1; r', \tau)Q^{-1}(r')\hat{H}^T(r')\, dr \qquad (6.35)$$

In view of (6.35), the term $T_1 T_1^*$ becomes

$$T_1 T_1^* = dk_0^{-1} \qquad (6.36)$$

where d is the real number defined by

$$d = \int_{t_0}^{t_1} \int_V \int_V [\hat{H}(r)G(r, t_1; r', \tau)B(r')R^{-1}B^T(\beta)Q^T(\beta)]$$
$$\times [G^*(\beta, t_1; r'', \tau)Q^{-1}(r'')\hat{H}^T(r'')]\, d\beta\, dr\, d\tau \qquad (6.37)$$

It follows from equations (6.35) and (6.36) that the pseudoinverse $T_1^*[T_1 T_1^*]^{-1}$ can be obtained in the form

$$T_1^+(\tau) = d^{-1} K_I(\tau) \tag{6.38}$$

where $K_I(\tau)$ is the M-dimensional column matrix defined by

$$K_I(\tau) = R^{-1} \int_V B^T(r) Q^T(r) G^*(r, t_1; \tau) Q^{-1}(r') \hat{H}^T(r') \, dr \tag{6.39}$$

6.3.1.3. Optimal Control: Solution to a Fredholm Integral Equation

Using the explicit relations for F^* and T_1^+, we can fully characterize the optimality condition (6.22). From expression (6.11) for T_1 and expression (6.27) for F^*FU, we obtain

$$T_1 F^* FU = k_0^{-1} \int_{t_0}^{t_1} K_{II}(\alpha) U(\alpha) \, d(\alpha) \tag{6.40}$$

where the kernel $K_{II}(\alpha)$ is defined by

$$K_{II}(\alpha) = \int_{t_0}^{t_1} \int_V \hat{H}(r) G(r, t_1; r', \tau) B(r') K(\tau; \alpha) \, dr \, d\tau \tag{6.41}$$

From (6.40), it follows that

$$T_1^+(\tau) T_1 F^* FU = (dk_0)^{-1} \int_{t_0}^{t_1} K_I(\tau) K_{II}(\alpha) U(\alpha) \, d\alpha \tag{6.42}$$

Using equations (6.42) and (6.27), we find that the optimality condition can be reformulated in terms of a nonhomogeneous Fredholm integral equation of the second kind,

$$U(\tau) = \Delta(\tau) + k_0^{-1} \int_{t_0}^{t_1} [d^{-1} K_I(\tau) K_{II}(\alpha) - K(\tau; \alpha)] U(\alpha) \, d\alpha \tag{6.43}$$

where $\Delta(\tau)$ is the function defined by equation (6.23).

It would be interesting to indicate the main differences between the problem treated in Chapter 4 and the problem treated here. Then solution to the optimal control problem in this chapter takes into account the additional requirement that the total power output should match a specified level p_1 at time t_1. If we compare equations (6.43) and (4.18), we see that there are two changes introduced in equation (4.18). The first is the addition of a new term $d^{-1} K_I(\tau) K_{II}(\tau)$ to the kernel $K(\tau; \alpha)$ of equation (4.18). The second effect is the addition of $T_1^+ \xi - T_1^+ T_1 F^* y$ to the forcing function $\Delta(\tau)$ in equation (4.18).

The advantages of the present optimization approach over more conventional techniques are self-evident. Techniques based on variational principles and modal expansion methods would yield necessary conditions for optimality in the form of an indefinite system of ordinary differential equations with mixed boundary conditions. In contrast, the necessary and sufficient condition (6.43) constitutes a finite system of integral equations, with as many equations as there are control devices in the reactor core.

Equation (6.43) is amenable to application of successive approximation techniques. As was the case in Chapter 4, a convergence analysis of the contraction mapping algorithm could be carried out with relative ease.

6.3.1.4. Practical Example

Consider the slab reactor model of the example given in Section 4.5. In the neighborhood of the equilibrium distribution $\phi_0(r)$,

$$\phi_0(r) = \phi_M \sqrt{\frac{2}{b}} \sin\left(\frac{\pi r}{b}\right) \tag{6.44}$$

the state deviation $\psi(r, t)$ at any time $t \geq t_0$ is

$$\psi(r, t) = G(r, t; r', t_0)Z_0(r') + F(r, t; \tau)U(\tau) \tag{6.45}$$

where

$$G(r, t; r', t_0)Z_0(r')$$
$$= \sum_n \frac{2}{b} \sin\left(\frac{n\pi r}{b}\right)\left[\int_0^b \sin\left(\frac{n\pi r'}{b}\right) Z_0(r') \, dr'\right] e^{\lambda_n(t-t_0)} \tag{6.46}$$

and

$$F(r, t; \tau)U(\tau) = \left[\frac{2}{b}\right]^{3/2} \phi_M V \int_{t_0}^t \sum_n e^{\lambda_n(t-\tau)} \sin\left(\frac{n\pi r}{b}\right)$$
$$\times \left[\sum_{i=1}^M \sin\left(\frac{n\pi r_i}{b}\right) \sin\left(\frac{\pi r_i}{b}\right) \mu_i(\tau)\right] d\tau \tag{6.47}$$

where $Z_0(r)$ is the state deviation at time t_0, and the eigenvalues λ_n are defined by

$$\lambda_n = \left[\frac{(v\Sigma_f - \Sigma_a)b^2}{\pi^2 D} - n^2\right] \frac{\pi^2 DV}{b^2} \tag{6.48}$$

Associated with $Z_0(r)$ there is a total power output p_0 given by

$$p_0 = e_f \Sigma_f \int_0^b Z_0(r) \, dr \tag{6.49}$$

If in addition to the requirement of the problem in Section 4.5 in which the control $U(\tau)$ minimizes

$$J(U) = \int_{t_0}^{t_1} \int_0^b \psi^2(r, t) \, dr \, dt + k_0 \int_{t_0}^{t_1} U^T(\tau)U(\tau) \, d\tau \tag{6.50}$$

the control function is also required to satisfy the power constraint

$$p_1 = 0 = e_f \Sigma_f \int_0^b \psi(r, t_1) \, dr \tag{6.51}$$

Then it follows from the previous discussion that the operators T_1, T_1^+, and T_1^* constitute the only new concepts required to fully characterize the necessary and sufficient condition for optimality. Extending these operators to the present case is given below.

In view of equations (6.51) and (6.11), it follows that

$$T_1 U = e_f \Sigma_f \int_0^b \int_{t_0}^{t_1} G(r, t_1; r', \tau) B(r') U(\tau) \, d\tau \, dr$$

$$= e_f \Sigma_f \int_0^b F(r, t_1; \tau) U(\tau) \, d\tau \tag{6.52}$$

Furthermore, from (6.47), we find that

$$T_1 U = e_f \Sigma_f \frac{4\phi_M V}{\pi} \left[\frac{2}{b}\right]^{1/2} \int_{t_0}^{t_1} \sum_{n=1}^{\infty} \sum_{i=1}^{M} \frac{\exp \lambda_{2n-1}(t_1 - \tau)}{2n - 1}$$

$$\times \sin\left(\frac{(2n - 1)\pi r_i}{b}\right) \sin\left(\frac{\pi r_i}{b}\right) u_i(\tau) \, d\tau \tag{6.53}$$

Recall from Section 4.5 that $B(r)$ denotes the M-dimensional row matrix operator with entries

$$V\phi_M \sqrt{\frac{2}{b}} \sin\left(\frac{\pi r}{b}\right) \delta(r - r_i) \tag{6.54}$$

Given that $G(r, t; r'\tau)$ is self-adjoint and in view of (6.35), it follows that the adjoint T_1^* is an M-dimensional, time-dependent column matrix with entries $T_1^*(\tau)$ given by

$$T_{1,i}^*(\tau) = e_f \Sigma_f \frac{4\phi_M V}{\pi k_0} \left[\frac{2}{b}\right]^{1/2} \sin\left(\frac{\pi r_i}{b}\right)$$

$$\times \sum_{n=1}^{\infty} \sin\left(\frac{(2n - 1)\pi r_i}{b}\right) \frac{\exp \lambda_{2n-1}(t_1 - \tau)}{2n - 1} \tag{6.55}$$

From equations (6.53) and (6.55), we find

$$T_1 T_1^* = \left[\frac{e_f \Sigma_f \phi_M V}{\pi}\right]^2 \frac{32}{b k_0}$$

$$\times \sum_{n=1}^{\infty} \sum_{m=1}^{\infty} \sum_{i=1}^{M} \sin\left(\frac{(2n - 1)\pi r_i}{b}\right) \sin\left(\frac{(2m - 1)\pi r_i}{b}\right) \sin^2\left(\frac{\pi r_i}{b}\right)$$

$$\times \frac{\exp(\lambda_{2n-1} + \lambda_{2m-1})(t_1 - t_0) - 1}{(2n - 1)(2m - 1)(\lambda_{2n-1} + \lambda_{2m-1})} \tag{6.56}$$

Finally, it follows from $T_1^+ = T_1^*[T_1 T_1^*]^{-1}$ that the pseudoinverse T_1^+ is an M-dimensional column matrix with entries $T_{1_i}^+(\tau)$ given by

$$T_{1_i}^+(\tau) = \frac{T_{1_i}^*(\tau)}{[T_1 T_1^*]} \tag{6.57}$$

It can be shown that the kernel $[d^{-1} K_I(\tau) K_{II}(\alpha) - K(\tau; \alpha)]$ in the optimality condition for this particular example is an M-dimensional matrix with entries $\bar{K}_{lj}(\tau; \alpha) - \hat{K}_{lj}(\tau; \alpha)$ in the lth row and jth column given by

$$\bar{K}_{lj}(\tau; \alpha) = C_0^{-1} \left[\frac{\phi_M 2V}{b} \right]$$

$$\times \sum_{p=1}^{\infty} \sum_{n=1}^{\infty} \sum_{m=1}^{\infty} \sum_{i=1}^{M} \exp[\lambda_{2p-1}(t_1 - \tau)$$

$$\times (2p - 1)^{-1}(2n - 1)^{-1}] \sin\left(\frac{(2n - 1)\pi r_i}{b}\right)$$

$$\times \sin^2\left(\frac{\pi r_i}{b}\right) \sin\left(\frac{(2p - 1)\pi r_i}{b}\right) \sin\left(\frac{\pi r_i}{b}\right) \sin\left(\frac{m \pi r_i}{b}\right)$$

$$\times \sin\left(\frac{m \pi r_j}{b}\right) \sin\left(\frac{\pi r_j}{b}\right) \int_{t_0}^{t_1} \exp\{\lambda_{2n-1}(t_1 - \beta)\} K_m(\beta; \alpha) \, d\beta \tag{6.58}$$

where

$$C_0 = \sum_{m=1}^{\infty} \sum_{m=1}^{\infty} \sum_{i=1}^{M} \sin\left(\frac{(2n - 1)\pi r_i}{b}\right) \sin\left(\frac{(2m - 1)\pi r_i}{b}\right) \sin^2\left(\frac{\pi r_i}{b}\right)$$

$$\times [\exp(\lambda_{2n-1} + \lambda_{2m-1})(t_1 - t_0) - 1]$$

$$\times (2n - 1)^{-1}(2m - 1)^{-1}(\lambda_{2n-1} + \lambda_{2m-1})^{-1} \tag{6.59}$$

and

$$K_m(\tau; \alpha) = \begin{cases} \exp \lambda_m(2t_1 - \tau - \alpha) - \exp \lambda_m |\tau \\ t_1 - \tau & \text{for } \lambda_m = 0 \text{ and } \alpha < \tau \\ t_1 - \alpha & \text{for } \lambda_m = 0 \text{ and } \alpha \geq \tau \end{cases} \tag{6.60}$$

Also,

$$\hat{K}_{lj}(\tau; \alpha) = \left[\frac{2\phi_M V}{b} \right]^2 \sum_{m=1}^{\infty} \sin\left(\frac{m \pi r_i}{b}\right) \sin\left(\frac{m \pi r_j}{b}\right)$$

$$\times \sin\left(\frac{\pi r_i}{b}\right) \sin\left(\frac{\pi r_j}{b}\right) K_m(\tau; \alpha) \tag{6.61}$$

Computing the optimal control requires that the kernel in the optimality condition be replaced by an approximate version, which, for example, could be obtained by truncating the series in the defining relations (6.58) and (6.61). In addition, the convergence rate of the successive approximation algorithm must be analyzed.

Although the present case is more involved, the derivation of error estimates between the true and the approximate kernels, and the computation of the convergence rate of the contraction mapping algorithm, could be carried out in the same manner as in Chapter 4.

6.4. Optimal Control of the State Distribution During Load-Following

A natural variation on the previous problem would arise if the requirement that the total power output from the reactor core match a specified level at time t_1 is replaced by the more stringent constraint that the total power output match a given load trajectory at a finite number of points in the time-interval of interest. This formulation clearly is relevant to the problem of controlling the reactor core during load-following operations. Although the previous and the present problems can be treated on an equal footing by the optimization technique of the minimum norm, the latter is more involved computationally.

Recall from the previous section that the relation connecting the control function U and the total power output $p(t_k)$ at time t_k is

$$\xi(t_k) = T(t_k; \tau) U(\tau) \tag{6.62}$$

where

$$\xi(t_k) = p(t_k) - \int_V \hat{H}(r) G(r, t_k; r', t_0) Z_0(r') \, dr \tag{6.63}$$

and $T(t_k; \tau)$ defines the transformation from H_1 into the real line:

$$T(t_k; \tau) = \int_V \hat{H}(r) F(r, t_k; \tau) U(\tau) \, dr$$

$$= \int_{t_0}^{t} \int_V \hat{H}(r) G(r, t_k; r', \tau) B(r') U(\tau) \, dr \, d\tau \tag{6.64}$$

It is clear that if the power output is specified at the finite set of points $\{t_k\}$, $k = 1, 2, \ldots, N_k$, along the time interval $[t_0, t_1]$, then the function $\xi(t_k)$ would also be completely specified in $\{t_k\}$.

A typical feature that characterize the optimization techniques of functional analysis is that the transformations involved in the formulation of the problem are defined on normed function spaces. Here the function $\xi(t_k), k = 1, 2, \ldots, N_k$, defined in (6.63) is an element of a finite-dimensional linear space. In fact, it could be considered to be an element of the Hilbert space H_4 (an N_k-dimensional Euclidean space) with inner product

$$\langle \xi, p \rangle_{H_4} = \sum_{k=1}^{N_k} \xi(t_k) p(t_k) \qquad (6.65)$$

If ξ denotes the column vector in H_4 with entries $\xi(t_k)$ defined by (6.62) and $\hat{T}$ represents the column matrix transformation with entries $T(t_k; \tau)$ defined by (6.64), then the present control problem could now be reformulated in the context of functional analysis: find the control $U \in H_1$ that minimizes

$$J(U) = \|FU - y\|_{H_2} + \|U\|_{H_1} \qquad (6.65a)$$

and satisfies

$$\xi = \hat{T} U \qquad (6.66)$$

Again, it follows from Porter's result that the necessary and sufficient conditions for optimality are

$$U = -F^* F U + \hat{T}^+ \eta + F^* y \qquad (6.67)$$

and

$$\xi = -\hat{T} F^* F U + \eta + \hat{T} F^* y \qquad (6.68)$$

where now η is an element of H_4 and $\hat{T}^+$ is the pseudoinverse of $\hat{T}$.

In this case the pseudoinverse involves the inversion of the N_k-dimensional matrix defined by $\hat{T}\hat{T}^*$, where, as usual, the asterisk denotes adjoint. For N_k small, the computation of $\hat{T}^+$ poses no problem, and conditions (6.67) and (6.68) may be combined to yield the operator equation

$$U = [\hat{T}^+ \hat{T} - I] F^* F U + \Delta \qquad (6.69)$$

where

$$\Delta = \hat{T}^+ \xi - [\hat{T}^+ \hat{T} - I] F^* y \qquad (6.70)$$

However, for large values of N_k the computation of $[\hat{T}\hat{T}^*]^{-1}$ becomes more difficult, and some suitable iterative method must be used. For this reason it is convenient to rewrite conditions (6.67) and (6.68) in a form that does not explicitly involve the pseudoinverse. Thus we define the function W in H_4 by

$$\hat{T}\hat{T}^* W = \eta \qquad (6.71)$$

Substituting (6.71) in (6.67) and (6.68), we obtain the desired result:

$$U = -F^* F U + \hat{T}^* W + F^* y \qquad (6.72)$$

and

$$0 = -\hat{T}F^*FU + \hat{T}\hat{T}^*W + \hat{T}F^*y - \xi \tag{6.73}$$

Finally, introducing the parameter ε in order to rewrite (6.73) in a form that lends itself to the application of successive iteration techniques, we have

$$W = \varepsilon\hat{T}F^*FU - [\varepsilon\hat{T}\hat{T}^* - I]W + \varepsilon[\xi - TF^*y] \tag{6.74}$$

in which ε may be used to improve the rate of convergence of the iterative algorithm employed.

In the limiting case wheren $N_k \to \infty$, the Hilbert space H_4 would no longer remain finite-dimensional. The inner product would be defined by

$$\langle \xi, p \rangle_{H_4} = \int_{t_0}^{t_1} \xi(t)p(t)\, dt \tag{6.75}$$

and the transformation $\hat{T}$ would simply become T, as defined in equation (6.47). The optimality conditions would then be given by expressions similar to equations (6.72) and (6.74), with T instead of $\hat{T}$. It is not difficult to show that in this limiting case the adjoint T^* is

$$T^*(\tau; t)\xi(t)$$

$$= k_0^{-1}\int_\tau^{t_1} R^{-1}\int_V B^T(r)Q^T(r)G^*(r, t; r', \tau)Q^{-1}(r')H^T(r')\, dr\, \xi(t)\, dt \tag{6.76}$$

We close this section with an important observation. The problem discussed above is based on the implicit assumption that the specified load trajectory lies in the range of T. This assumption is required in order to ensure that the equation

$$TU = \xi \tag{6.77}$$

has at least one solution.

6.5. Optimal Control of the State Distribution with Fixed End State

Another variation on the problem of controlling the state distribution along a given time interval would arise if, in addition to the requirement that the control function U minimize the functional

$$J(U) = \|FU - y\|_{H_2} + \|U\|_{H_1} \tag{6.78}$$

we also consider the requirement that the state reach a specified distribution $Z_1(r)$ at time t_1. It is clear that this formulation would contain the case treated in Section 6.3, in which only the total power output is required to match a given power level at the end of the time interval. The present problem, however, would be much more restrictive and consequently much more involved computationally.

The additional restriction imposed on U would then have the form

$$Z_1(r) - G(r, t_1; r', t_0)Z_0 = F(r, t_1; \tau)U(\tau) \tag{6.79}$$

where all the variables are defined as before. The optimality conditions would be

$$U = -F^*FU + F_1^*W + F^*y \tag{6.80}$$

and

$$W = \varepsilon F_1 F^*FU - [\varepsilon F_1 F_1^* - I]W + \varepsilon[\xi - F_1 F^*y] \tag{6.81}$$

where F_1 denotes the operator $F(r, t_1; \tau)$ mapping H_1 into H, W is an unknown function in H, and ξ is a function defined by

$$\xi(r) = Z_1(r) - G(r, t_1; r', t_0)Z_0 \tag{6.82}$$

The adjoint F_1^* is

$$F_1^*(\tau; r')W(r') = k_0^{-1}R^{-1}\int_V B^T(r)Q^T(r)G^*(r, t_1; r', \tau)W(r') \, dr \tag{6.83}$$

and the other variables in equations (6.79) and (6.81) are defined as before.

Although conditions (6.80) and (6.81) are amenable to the application of successive iteration techniques, the degree of complexity posed by these conditions is formidable. To show how difficult it would be to implement an iterative algorithm, it suffices to realize that at each iteration step, in addition to the trial control function, at least one state distribution defined throughout the reactor core and the time interval $[t_0, t_1]$ would have to be stored in memory. This requirement severely limits the application of the present problem formulation.

The degree of computational complexity that characterizes this problem could be reduced considerably if the requirement that the state of the reactor core match a specified distribution at time t_1 is replaced by the less restrictive requirement that only a projection of the state match a given distribution in a finite-dimensional subspace of H at time t_1. Specifically, if the state distribution $\psi(r, t_1)$ is represented in terms of a complete basis $\{\psi_n\}$ in H,

$$\psi(r, t_1) = \sum_{n=1}^{\infty} x_n(t_1)\psi_n(r) \tag{6.84}$$

where $\{x_n\}$ is the set of expansion coefficients, then the formulation of the less restrictive requirement is

$$x_n(t_1) - \langle G(t_1; t_0)Z_0, \psi_n^* \rangle_H = \langle F(t_1; \tau)U(\tau), \psi_n^* \rangle_H \qquad (6.85)$$

$$n = 1, 2, \ldots, M_1$$

where $\{\psi_n^*\}$ is the associated dual set of $\{\psi_n\}$, satisfying

$$\langle \psi_m, \psi_n^* \rangle = \delta_{mn} \qquad (6.86)$$

and M_1 is the dimension of the subspace of H into which the state is projected.

It is clear that in the limit as $M_1 \to \infty$ the requirement (6.85) would become identical to (6.79). For the particular case, for example, where $\{\psi_n\}$ and $\{\psi_n^*\}$ are chosen to be the natural modes, described by Kaplan (Ref. 3.8), and that satisfy the relations

$$A\psi_n = \lambda_n \psi_n \qquad (6.87)$$

and

$$A^* \psi_n^* = \lambda_n \psi_n^* \qquad (6.88)$$

where A is the operator that appears in the state equation

$$\frac{\partial \psi}{\partial t}(r, t) = A(r)\psi(r, t) + B(r)U(t) \qquad (6.89)$$

the coefficients $x_n(t)$ of the expansion

$$\psi(r, t) = \sum_{n=1}^{\infty} x_n(t)\psi_n(r) \qquad (6.90)$$

are

$$x_n(t) = e^{\lambda_n(t-t_0)} \langle Z_0, \psi_n^* \rangle_H + \sum_{j=1}^{M} \int_{t_0}^{t} e^{\lambda_n(t-\tau)} \langle B_j, \psi_n^* \rangle_H u_j(\tau)\, d\tau \quad (6.91)$$

where B_j is the jth column of B and u_j is the ith entry in the control vector U. Also, the linear constraint in (6.85) would take the form

$$x_{1_n} - e^{\lambda_n(t_1-t_0)} \langle Z_0, \psi_n^* \rangle_H = \sum_{j=1}^{M} \int_{t_0}^{t_1} e^{\lambda_n(t-\tau)} \langle B_j, \psi_n^* \rangle_H u_j(\tau)\, d\tau \quad (6.92)$$

$$n = 1, 2, \ldots, M_1$$

where the $\{x_{1_n}\}$ are the coefficients in the series representation of $Z_1(r)$.

6.5.1. Suboptimal Control

So far we have not considered problems in which the space of controls is finite-dimensional. This situation arises naturally when digital computers constitute the main controller in a feedback system, or artificially when either time-discretization or function–expansion techniques are applied in order to simplify an optimal control problem.

In what follows, it is assumed that the control function is given by a finite series expansion in terms of a set of known functions of time $\omega_j(t)$, $j = 1, 2, \ldots, M_2$; that is,

$$u_i(t) = \sum_{j=1}^{M_2} \alpha_{ij}\omega_j(t) \tag{6.93}$$

where $\{\alpha_{ij}\}$ are the expansion coefficients.

This concept applies equally well when the control function is discrete or continuous. The specific application depends on the nature of the functions $\{\omega_j\}$.

In view of the modal expansions (6.90) and (6.93), it follows that the cost function (6.8) can be written in the form

$$J(U) = \int_{t_0}^{t_1} \sum_{j=1}^{M} \sum_{m=1}^{M} (x_j(t) - z_j(t))\rho_{jm}(x_m(t) - z_m(t)) \, dt$$

$$+ k_0 \sum_{n=1}^{M} \sum_{i=1}^{M} \sum_{j=1}^{M_2} \sum_{k=1}^{M_2} \alpha_{nj}\beta_{njki}\alpha_{ik} \tag{6.94}$$

where $z_j(t)$ is the jth expansion coefficient of the desired distribution $Z(r, t)$, ρ_{jm} is given by

$$\rho_{jm} = \int_V \psi_j^{\mathrm{T}}(r)Q(r)\psi_m(r) \, dr \tag{6.95}$$

and

$$\beta_{njki} = \int_{t_0}^{t_1} \omega_j(t)r_{ni}\omega_k(t) \, dt \tag{6.96}$$

where r_{ni} is the nith entry in the matrix R. Similarly, substituting equation (6.93) in (6.91), we find that the expansion coefficients x_n are

$$x_n(t) = e^{\lambda_n(t-t_0)}\langle Z_0, \psi_n^* \rangle_H + \sum_{j=1}^{M} \sum_{k=1}^{M_2} \int_{t_0}^{t} e^{\lambda_n(t-\tau)}\langle \beta_j, \psi_n^* \rangle_H \omega_k(\tau) \, d\tau \, \alpha_{jk} \tag{6.97}$$

which for $n = 1, 2, \ldots, M_1$ and evaluated at t_1 also define the linear constraints (6.92).

To reformulate the optimal control problem in the context of functional analysis, we introduce the following function spaces: H_s shall represent the Hilbert space of time-dependent sequences, with inner product

$$\langle x, y \rangle_{H_s} = \int_{t_0}^{t_1} \sum_{n=1}^{\infty} \sum_{j=1}^{\infty} y_j(t)\rho_{nj}x_n(t) \, dt \tag{6.98}$$

where the coefficients ρ_{nj} are defined in (6.95). H_c will denote the finite-dimensional Hilbert space of real matrices with inner product

$$\langle \alpha, \phi \rangle_{H_c} = k_0 \sum_{j=1}^{M_2} \sum_{k=1}^{M_2} \sum_{n=1}^{M} \sum_{i=1}^{M} \alpha_{nj} \beta_{njki} \phi_{ik} \qquad (6.99)$$

where the coefficients β_{njki} are defined in (6.96), and the variables α_{nj} and ϕ_{ik} denote the njth and ikth entries in α and ϕ, respectively. Finally, H_T represents the finite-dimensional Hilbert space with inner product

$$\langle \xi, \gamma \rangle_{H_T} = \sum_{i=1}^{M_1} \xi_i \gamma_i \qquad (6.100)$$

6.5.2. The Minimum Norm Formulation

Using all the above definitions, we can formulate the suboptimal control problem as follows: find the control element α in H_c that minimizes

$$J(U) = \|\tilde{F}\alpha - \hat{y}\|_{H_s} + \|\alpha\|_{H_c} \qquad (6.101)$$

subject to the linear constraint

$$\xi = \tilde{T}\alpha \qquad (6.102)$$

where $\tilde{F}$ is a linear transformation from H_c into H_s with entries defined by

$$\tilde{F}_n(t)\alpha = \sum_{j=1}^{M} \sum_{k=1}^{M_2} g_{nk}(t)\langle \beta_j, \psi_n^* \rangle_H \alpha_{jk}; \qquad n = 1, 2, \ldots . \qquad (6.103)$$

where

$$g_{nk}(t) = \int_{t_0}^{t} e^{\lambda_n(t-\tau)} \omega_k(\tau) \, d\tau \qquad (6.104)$$

$\tilde{T}$ is a linear transformation from H_c into H_T with entries $\tilde{F}_n(t_1)$, $n = 1, 2, \ldots, M_1$, and ξ is an element of H_T given by

$$\xi_n = x_{1_n} - e^{\lambda_n(t_1-t_0)}\langle Z_0, \psi_n^* \rangle H \qquad (6.105)$$

and $\hat{y}$ is a function defined in H_s by

$$\hat{y}_n(t) = z_n(t) \, e^{\lambda_n(t-t_0)}\langle Z_0, \psi_n^* \rangle_H \qquad (6.106)$$

From Porter's result it follows that the unique solution to this problem is

$$\alpha = [I + \tilde{F}^*\tilde{F}]^{-1}[\tilde{T}^+\eta + \tilde{F}^*y] \qquad (6.107)$$

where η is the unique element of H_T satisfying

$$\xi = \tilde{T}[I + \tilde{F}^*\tilde{F}]^{-1}[\tilde{T}^+\eta + \tilde{F}^*y] \qquad (6.108)$$

The adjoint $\tilde{F}^*$ is defined by the inner-product relation

$$\langle \tilde{F}\alpha, y \rangle_{H_c} = \langle \alpha, \tilde{F}^* y \rangle_{H_c} \tag{6.109}$$

Substituting equation (6.103) in the left side of equation (6.109), we find that

$$\langle \tilde{F}\alpha, y \rangle_{H_c} = \int_{t_0}^{t_1} \sum_{n=1}^{\infty} \sum_{i=1}^{\infty} \sum_{j=1}^{M} \sum_{k=1}^{M_2} y_i(t) \rho_{ni} \langle \beta_j, \psi_n^* \rangle_H g_{nk}(t) \alpha_{jk} \tag{6.110}$$

Introducing a new index p that uniquely corresponds to pairs (j, k), we rewrite equation (6.110) in the form

$$\langle \tilde{F}\alpha, y \rangle \sum_{p=1}^{M_3} \phi_p(y) \alpha_p \tag{6.111}$$

where $M_3 = M \times M_2$ and ϕ_p is a functional in H_s defined by

$$\phi_p(y) = \int_{t_0}^{t_1} \sum_{n=1}^{\infty} \sum_{i=1}^{\infty} \langle \beta_j, \psi_n^* \rangle_H g_{nk}(t) \rho_{ni} y_i(t) \, dt \tag{6.112}$$

In view of equation (6.111), we find by inspection that

$$\tilde{F}^* y = k_0^{-1} [\boldsymbol{\phi}(y) \quad \beta^{-1}]^T \tag{6.113}$$

where β is the positive-definite matrix of order M_3 with entries defined in equation (6.96), and $\boldsymbol{\phi}$ is a column transformation from H_s into H_c with entries ϕ_p given in equation (6.112).

Similarly, it can be shown that the adjoint $\tilde{T}^*$ is

$$\tilde{T}^* \xi = k_0^{-1} \beta^{-1} \boldsymbol{\theta} \xi \tag{6.114}$$

where $\boldsymbol{\theta}$ is an $M_3 \times M_1$ matrix with entries

$$\theta_{pn} = \langle B_j, \psi_n^* \rangle_H g_{nk}(t_1) \tag{6.115}$$

in which p uniquely corresponds to the pair (j, k).

Finally, it follows from equations (6.113), (6.114), and (6.103) that the operator $[\tilde{F}^* \tilde{F}]$ is a matrix of dimension M_3 defined by

$$\tilde{F}^* \tilde{F} = k_0^{-1} \beta^{-1} \hat{F} \tag{6.116}$$

where $\hat{F}$ is a positive-definite matrix with entries given by

$$\hat{f}_{qp} = \sum_{n=1}^{\infty} \sum_{i=1}^{\infty} \langle b_j, \psi_n^* \rangle_H \langle B_r, \psi_i^* \rangle_H \hat{g}_{nmik} \tag{6.117}$$

where

$$\hat{g}_{nmik} = \int_{t_0}^{t_1} \rho_{ni} g_{im}(t) g_{nk}(t) \, dt \tag{6.118}$$

and the indices q and p correspond uniquely to pairs (r, m) and (j, k), respectively.

6.5.3. Practical Example

Consider the same example mentioned in Section 4.5, which is the homogeneous slab reactor model described by the equation

$$\frac{\partial \psi}{\partial t}(r, t) = VD\frac{\partial^2}{\partial r^2} \psi(r, t) + V[v\Sigma_f - \Sigma_a]\psi(r, t)$$

$$- V \sum_{i=1}^{M} u_i(t)\phi_0(r)\delta(r - r_i) \qquad (6.119)$$

with boundary condition

$$\psi(0, t) = \psi(b, t) = 0 \qquad (6.120)$$

and steady-state distribution

$$\phi_0(r) = \phi_M \sqrt{\frac{2}{b}} \sin\left(\frac{\pi r}{b}\right) \qquad (6.121)$$

In this case the operator $VD(\partial^2/\partial r^2) + V[v\Sigma_f - \Sigma_a]$ is self-adjoint and generates the orthonormal eigenfunctions

$$\psi_n(r) = \psi_n^*(r) = \sqrt{\frac{2}{b}} \sin\left(\frac{n\pi}{b} r\right) \qquad (6.122)$$

and eigenvalues

$$\lambda_n = \left[\frac{v\Sigma_f - \Sigma_a}{\pi^2 D} b^2 - n^2\right] V\frac{\pi^2}{2} D \qquad (6.123)$$

Also, for simplicity, Q and R are assumed to be the identity matrices. The desired trajectory $Z(r, t)$ and the desired final distribution $Z_1(r)$ are both chosen to be zero.

$$Z(r, t) = Z_1(r) = 0 \qquad (6.124)$$

If the functions $\{\omega_j(t)\}$ are chosen to be the unit pulses

$$\omega_j(t) = \begin{cases} 1, & \tau_{j-1} < t \leq \tau_j \\ 0, & \text{elsewhere} \end{cases} \qquad (6.125)$$

where

$$\tau_j = \frac{t_1 - t_0}{M_2} j, \qquad j = 1, \ldots, M_2 \qquad (6.126)$$

then the coefficients of the modal expansion (6.91) satisfy

$$x_n(t) = e^{\lambda_n(t-t_0)}\langle Z_0, \psi_n\rangle_H + \sum_{j=1}^{M} \sum_{k=1}^{M_2} g_{nk}(t)\langle B_j, \psi_n\rangle_H \alpha_{jk} \qquad (6.127)$$

where

$$\langle B_j, \psi_n \rangle = -\sqrt{\frac{2}{b}} \, \phi_0(r_j) \, V \sin\left(\frac{n\pi}{b} r_j\right) \tag{6.128}$$

and

$$g_{nk}(t) = \int_{t_0}^t e^{\lambda_n(t-\tau)} \omega_k(\tau) \, d\tau \tag{6.129}$$

which for $\lambda_n = 0$ takes on values

$$g_{nk}(t) = \begin{cases} 0, & t \le \tau_{k-1} \\ t - \tau_{k-1}, & \tau_{k-1} < t \le \tau_k \\ \dfrac{t_1 - t_0}{M_2}, & \tau_k < k \end{cases} \tag{6.130}$$

and for $\lambda_n \ne 0$,

$$g_{nk}(t) = \begin{cases} 0, & t \le \tau_{k-1} \\ \lambda_n^{-1}[\exp\{\lambda_n(t - \tau_{k-1}) - 1\}, & \tau_{k-1} < t \le \tau_k \\ \lambda_n^{-1}[e^{\lambda_n(t-\tau_{k-1})} - e^{\lambda_n(t-\tau_k)}], & \tau_k < t \end{cases} \tag{6.131}$$

The adjoint $\tilde{F}^*$ is defined by

$$F^* y = k_0^{-1} \phi(y) \tag{6.132}$$

where

$$\phi_p(y) = -\sum_{n=1} \langle B_j, \psi_n \rangle_H \int_{t_0}^{t_1} g_{nk}(t) \, e^{\lambda_n(t-t_0)} \, dt \, \langle Z_0, \psi_n \rangle_H \tag{6.133}$$

and the index p is

$$p = M_2(j - 1) + k \tag{6.134}$$

Also, the matrix $\tilde{F}^* \tilde{F}$ is defined by

$$\tilde{F}^* \tilde{F} = k_0^{-1} \hat{F} \tag{6.135}$$

where $\hat{F}$ denotes a matrix with entries

$$\hat{f}_{pq} = \sum_{n=1} \langle B_i, \psi_n \rangle_H \langle B_j, \psi_n \rangle_H \hat{g}_{nmnk} \tag{6.136}$$

and

$$\hat{g}_{nmnk} = \int_{t_0}^{t_1} g_{mn}(t) g_{nk}(t) \, dt \tag{6.137}$$

The indices p and q in (6.127) are determined uniquely from

$$q = M_2(j - 1) + m \tag{6.138}$$

$$p = M_2(i - 1) + k \tag{6.139}$$

where m and k take on integer values in $[1, M_2]$ and j and i take on integer values in $[1, M]$.

The adjoint $\tilde{T}^*$ is

$$\tilde{T}^* = k_0^{-1}\theta \tag{6.140}$$

where θ is a rectangular matrix with entries

$$\theta_{pn} = \langle B_j, \psi_n \rangle_H g_{nk}(t_1) \tag{6.141}$$

Also,

$$\xi_n = -e^{\lambda_n(t_1 - t_0)}\langle Z_0, \psi_n \rangle_H \tag{6.142}$$

The data for the slab reactor model are given by Wiberg in Table 6.1. In this case, the eigenvalues are real. The first eigenvalue is close to zero, and all other eigenvalues are negative. Four cases have been studied as follows:

1. *Case 1.*

Figure 6.1 shows the trajectory of the flux deviations along the time interval $[0.0, 2.5]$ sec when there is no control action applied. At the end of the time interval all the higher harmonics have decayed almost completely, and only the fundamental mode remains virtually unchanged. In all cases, the control functions considered belong to the finite-dimensional space spanned by pulse functions of duration 0.25 sec.

Figure 6.2 shows the trajectory of the flux deviations when the control action minimizes the performance index,

$$J(U) = \int_{t_0}^{t_1} \int_0^b \psi^2(r, t)\, dr\, dt + k_0 \int_{t_0}^{t_1} \sum_{i=1}^{2} u_i(t)\, dt \tag{6.143}$$

Table 6.1. Data For Numerical Examples

Migration length	$D/\Sigma_a = 160 \text{ cm}^2$
Average neutron lifetime	$[\Sigma_a V]^{-1} = 0.1 \text{ sec}$
Infinite homogeneous multiplication constant	$v\Sigma_f/\Sigma_a = 1.0256$
Neutrons/fission	$v = 2.5$
Diffusion coefficient	$D = 0.5070 \text{ cm}$
Reactor width	$b = 250 \text{ cm}$
Time interval $[t_0, t_1]$	$t_0 = 0.0 \text{ sec}$
	$t_1 = 2.5 \text{ sec}$
Weighting parameter	$k_0 = 10.000$
Number of rods	$M = 2$
Number of linear constraints in equation (6.142)	$M_1 = 3$
Duration of pulses $\omega_k(t)$	$[t_1 - t_0]/M_2 = 0.25 \text{ sec}$
Control rod location	$r_1 = 66 \text{ cm}, r_2 = 150 \text{ cm}$
Initial state distribution	$Z_0(r) = \phi_M[0.05\psi_1(r) + 0.03\psi_2(r) + 0.01\psi_3(r)]$
Desired state distribution	$Z(r) = 0.0$

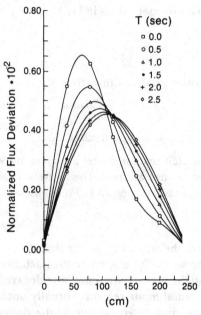

Figure 6.1. The uncontrolled flux distribution.

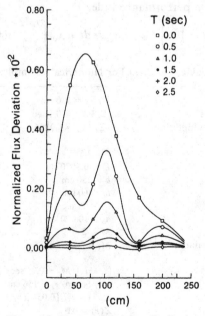

Figure 6.2. The controlled flux distribution.

and satisfies the constraint that the fundamental and the first two harmonics vanish at the end of the time-interval. The optimal control is given in Table 6.2.

2. *Case 2.*

Figure 6.3 shows the trajectory of the flux deviations when the control function is only required to minimize the cost functional in equation (6.143). Although the results of Figures 6.2 and 6.3 are almost identical, it is clearly evident from Figure 6.3 that at the end of the time interval the fundamental mode is still present. The optimal control for this case is given in Table 6.2.

3. *Case 3.*

Figure 6.4 shows the trajectory of the flux deviation when the control is required to satisfy the constraint that the fundamental and the two first harmonics vanish at the end of the time interval and minimize the performance index

$$J(U) = \int_{t_0}^{t_1} \sum_{i=1}^{2} u_i(t) \tag{6.144}$$

The effect of the optimal control function upon the trajectory compares poorly against the first two cases, as should be expected. The optimal control function computed in this case constitutes the effortless way of satisfying the given constraint on the final state. The control function is given in Table 6.3.

Table 6.2. Optimal Control Functions For Figures 6.1 and 6.2

	Case 1		Case 2	
t (sec)	rod at $r = 66$ cm $u_1(t)$	rod at $r = 150$ cm $u_2(t)$	rod at $r = 66$ cm $u_1(t)$	rod at $r = 150$ cm $u_2(t)$
0.00	0.274×10^{-2}	0.139×10^{-2}	0.274×10^{-2}	0.139×10^{-2}
0.25	0.168×10^{-2}	0.915×10^{-3}	0.168×10^{-2}	0.918×10^{-3}
0.50	0.105×10^{-2}	0.618×10^{-3}	0.105×10^{-2}	0.621×10^{-3}
0.75	0.673×10^{-3}	0.426×10^{-3}	0.664×10^{-3}	0.420×10^{-3}
1.00	0.432×10^{-3}	0.289×10^{-3}	0.420×10^{-3}	0.282×10^{-3}
1.25	0.271×10^{-3}	0.189×10^{-3}	0.265×10^{-3}	0.188×10^{-3}
1.50	0.176×10^{-3}	0.127×10^{-3}	0.165×10^{-3}	0.122×10^{-3}
1.75	0.124×10^{-3}	0.948×10^{-4}	0.982×10^{-4}	0.761×10^{-4}
2.00	0.850×10^{-4}	0.714×10^{-4}	0.520×10^{-4}	0.132×10^{-4}
2.25	0.505×10^{-4}	0.569×10^{-4}	0.162×10^{-4}	0.132×10^{-4}

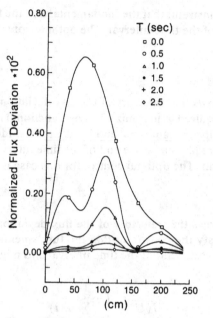

Figure 6.3. The controlled flux distribution.

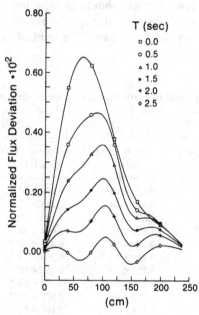

Figure 6.4. The controlled flux distribution.

Table 6.3. Optimal Control Functions For Figures 6.3 and 6.4

	Case 3		Case 4	
t (sec)	rod at $r = 66$ cm $u_1(t)$	rod at $r = 150$ cm $u_2(t)$	rod at $r = 66$ cm $u_1(t)$	rod at $r = 150$ cm $u_2(t)$
0.00	0.593×10^{-3}	0.578×10^{-3}	0.565×10^{-3}	0.183×10^{-4}
0.25	0.497×10^{-3}	0.481×10^{-3}	0.118×10^{-2}	0.663×10^{-3}
0.50	0.500×10^{-3}	0.478×10^{-3}	0.115×10^{-2}	0.676×10^{-3}
0.75	0.600×10^{-3}	0.574×10^{-3}	0.482×10^{-3}	0.551×10^{-4}
1.00	0.606×10^{-3}	0.571×10^{-3}	0.443×10^{-3}	0.724×10^{-3}
1.25	0.517×10^{-3}	0.469×10^{-3}	0.103×10^{-2}	0.728×10^{-3}
1.50	0.526×10^{-3}	0.463×10^{-3}	0.975×10^{-3}	$0.749 \times 1t^{-3}$
1.75	0.640×10^{-3}	0.550×10^{-3}	0.270×10^{-3}	0.133×10^{-3}
2.00	0.662×10^{-3}	0.535×10^{-3}	0.173×10^{-3}	0.132×10^{-3}
2.25	0.600×10^{-3}	0.414×10^{-3}	0.578×10^{-3}	0.641×10^{-3}

4. Case 4

Figure 6.5 shows the flux deviations for the case where the control function minimzes the performance index

$$J(U) = \int_0^b \psi^2(r, t_1) \, dr + k_0 \int_{t_0}^{t_1} \sum_{i=1}^{2} u_i(t) \, dt \qquad (6.145)$$

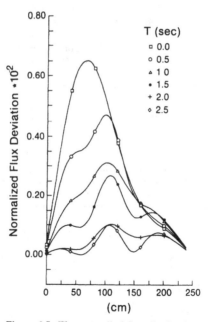

Figure 6.5. The controlled flux distribution.

The performance of the optimal control in this case does not compare favorably against the first two cases, in which the deviations of the flux are penalized along the trajectory. Although this result should be expected, the comparison of the present against the first two cases is not entirely valid, since for the same weighting parameter k_0 in the performance indices in equations (6.143) and (6.145) the absence of the time integral in the first term of (6.145) implicitly gives more weight to the second term of the performance index, which penalizes the control effort. Consequently, the norm of the optimal control in this case is almost two times smaller than the corresponding norms in the first two cases. The optimal control is given in Table 6.3.

References

6.1. AXELBAND, E. I., and KREIN, M., *Some Questions in the Theory of Moments*, American Mathematical Society, Providence, Rhode Island (translated from the 1938 Russian Edition), 1962.

6.2. AZIZ, A. K., WINGATE, J. W., and BALAS, M. J. (Editors), *Control Theory of Systems Governed by Partial Differential Equations*, Academic Press, New York, 1977.

6.3. ASSATANI, K., "Near Optimum Control of Distributed Parameter Systems Via Singular Perturbation Theory," *J. Math. Appl.* **54**, 799–810 (1976).

6.4. AXELBAND, E. I., "Optimal Control of Linear Distributed Parameter Systems," in *Adv. Control Syst.*, edited by C. T. Leondes, Vol. 7, pp. 257–300, 1969.

6.5. BOBONE, R., "A Computerized Solution of the Diffusion Equations by the Method of Solution Functions," *Trans. Am. Nucl. Soc.* **9**, 473–485 (1966).

6.6. BOBONE, R., "The Method of Solution Functions Extended from $r - \sigma$ to $r - z$ and $r - \sigma - z$ Geometry," *Trans. Am. Nucl.* **10**, 548–557 (1967).

6.7. BOBONE, R., "The Method of Solution Functions: A Computer-Oriented Solution of Boundary Value Problems as Applied to Nuclear Reactors," *Nucl. Sci. Eng.* **29**, 337–345 (1967).

6.8. BUTKOVSKIY, A. G., "Control in Distributed Systems (A Review)," Proceedings of the IFAC Symposium on the control of Distributed Parameter Systems, edited by M. H. Hamza, Banff, Canada, 1971.

6.9. BUTKOVSKIY, A. G., EGOROV, A. I., and LURIE, K. A., "Optimal Control of Distributed Systems," *SIAM Journal of Control* **6**(3), 437–476 (1968).

6.10. BUTKOVSKIY, A. G., *Distribution Control Systems*, American Elsevier, New York, 1969.

6.11. BUTKOVSKIY, A. G., "The Method of Moments in the Theory of Optimal Control of Systems with Distributed Parameter Systems," *Automation and Remote Control* **24**, 1106–1125 (1964).

6.12. BANACH, S., *Theorie des Operations lineaires*, Chelsea, New York, 1932.

6.13. BALAKRISHNAN, A. V., "An Operator Theoretic Formulation of a Class of Control Problems and Steepest Descent Method of Solution," *SIAM J. Control* **1**(2), 109–127 (1963).

6.14. BALAKRISHNAN, A. V., "Optimal Control Problems in Banach Spaces," *J. SIAM Control* **3**, 152–165 (1965).

6.15. COTTON, F. A., *Chemical Applications of Group Theory*, Wiley-Interscience, New York, 1971.

6.16. CHAUDHURI, S. P., "Distributed Optimal Control in a Nuclear Reactor," *Int. J. Control* **16**(5), 927–935 (1972).

6.17. DETTMAN, J. W., *Mathematical Methods in Physics and Engineering*, McGraw-Hill, New York, 1969.

6.18. EL-WAKIL, M. M., *Nuclear Power Engineering*, McGraw-Hill, New York, 1962.

6.19. EL-WAKIL, M. M., *Nuclear Energy Conversion*, International, Scranton, 1971.

6.20. EL-WAKIL, M. M., *Nuclear Heat Transport*, International, Scranton, 1971.

6.21. GRUMBACH, R., "On-Line Control of the Neutron Flux Distribution in a Nuclear Reactor Core," Proceedings of the IFAC 5th World Congress, Paper No. 6.2, Paris, 1972.

6.22. GREEN, C. D, *Integral Equation Methods*, Barnes and Noble, New York, 1969.

6.23. HABIB, I. S., *Engineering Analysis Methods*, Lexington Books, Toronto, 1975.

6.24. HOCHSTADT, H., *Integral Equations*, Wiley-Interscience, New York, 1973.

6.25. HELLY, E., "Uber Lineare Funktionaloperationen," *Wiener Berichte* **121**, 265–267 (1912).

6.26. HILLE, E., and PHILLIPS, R. S., *Functional Analysis and Semigroups*, American Mathematical Society, Providence, Rhode Island, 1957.

6.27. HAMZA, M. H., and RASMY, M. E., "Optimal Control of Distributed Parameter Systems with Discrete Constrained Inputs," *Int. J. Control* **20**(1), 159–170 (1974).

6.28. HAMERMESH, M., *Group Theory*, Addison-Wesley, Reading, Massachusetts, 1962.

6.29. ISBIN, H. S., *Introductory Nuclear Reactory Theory*, Reinhold, New York, 1963.

6.30. KANTOROVICH, L. V., and KRYLOV, V. I., *Approximate Methods of Higher Analysis*, Noordhoff, Groningen, 1958.

6.31. KREYSZIG, E., *Introductory Functional Analysis with Applications*, Wiley, New York, 1978.

6.32. KRASOVSKII, N. N., "On the Theory of Optimal Regulation," *Aut. and Remote Control* **18**, 1005–1022 (1957).

6.33. KLIGER, I., "Optimal Control of a Space-Dependent Nuclear Reactor," *Trans. Am. Nucl. Soc.* **8**, 233–245 (1965).

6.34. KURODA, Y., and MAKINO, A., "Optimal Control for a Class of Nuclear Reactor Systems with Distributed Parameters," Proceedings of the IFAC Symposium on the Control of Distributed Parameter Systems, Vol. II, Paper No. 10-1, Banff, Canada, 1971.

6.35. KHANA, M., PAUKSENS, J., and HINCHLEY, E. M., "The Use of Flux Mapping in the On-Line Spatial Control of the 600 M(c)Candu-PHW Reactor," Joint Aut. Control Conf. Vol. 1, pp. 375–387 (1977).

6.36. LUENBERGER, D. G., *Optimization by Vector Space Methods*, Wiley, New York, 1969.

6.37. LEWINS, J., and BABB, A. L., "Optimum Nuclear Reactor Control Theory", *Adv. in Nuc. Sc. and Technology* **4**, 252–260 (1968).

6.38. LOMONT, J. S., *Applications of Finite Groups*, Academic Press, New York, 1959.

6.39. MIKHLIN, S. G., *Integral Equations*, Pergamon Press, New York, 1964.

6.40. MITTER, S. K., "Optimal Control of Distributed Parameter Systems", Control of Distributed Parameter Systems, 1969 Joint Automatic Control Conference, Boulder, Colorado, August 6, 1969; ASME, New York, 13, 1969.

6.41. MIZUKAMI, K., "Applications of Functional Analysis to Optimal Control Problems," *Control Theory and Topics in Functional Analysis*, International Atomic Energy Agency, Vol. 2, pp. 39–73, Vienna, 1976.

6.42. NIEVA, R., and CHRISTENSEN, G. S., "Symmetry Reduction of Reactor Systems," *Nucl. Sci. Eng.* **64**, 791–795 (1977).

6.43. NIEVA, R., and CHRISTENSEN, G. S., "Symmetry Reduction of Linear Distributed Parameter Systems," *Int. J. Control* **36**(1), 143–153 (1982).

7

Conclusions

7.1. Summary

The main objective for the reactor control system is to maintain the neutron flux distribution within acceptable limits in order to avoid potentially harmful spots with high power density. Optimal control theory can play an important role in improving the performance of existing reactor control systems, and considerable research has been done in this area.

In Chapter 2, we reviewed some of the basic optimization methods. We began with matrix operations and considered, in turn, (1) calculus of variations, (2) dynamic programming and the principle of optimality, (3) Pontryagin's maximum principle, and (4) some minimum norm problems from functional analysis.

Modal expansion and modal decomposition techniques and their applications to distributed parameter systems were discussed in Chapter 3. Decomposition according to widely different time constants and model reduction by means of symmetry reduction considerations were discussed. We reviewed the rudiments of the representation theory of groups and symmetry principles.

Chapter 4 dealt with the optimal control of distributed nuclear reactors. A variety of problems was considered. Generally, we change the state of the reactor into the vicinity of a desired state distribution while minimizing a performance index that penalizes both the control effort and the deviations of the reactor state from the desired distribution.

Chapter 5 addressed different techniques used to control the nuclear reactor during load-following operations. We discussed applications of multistage mathematical programming linear-quadratic programming, and multilevel methods.

Chapter 6 was devoted to applications of minimum norm optimization techniques to the problem of controlling the shape and intensity of the power density distribution.

Application of the minimum norm approach to all problems in this book was described in terms of a general linear distributed parameter model. Several numerical examples were presented.

7.2. Future Work

The application of the minimum norm technique to situations in which the saturation of the control functions is also taken into account constitutes a natural extension to the present work. Although in theory this type of problem can be formulated and treated in the context of functional analysis, the implementation of the optimizing algorithms is difficult. More work is needed in this area.

An ideal approach to the optimal control of linear distributed parameter systems should combine the computational advantages of the methods of functional analysis with the engineering benefits of the feedback control concept, which characterizes the Kalman-Pontryagin theory of optimal control for lumped linear systems. It would be worthwhile to investigate, in this spirit, the feasiblity of synthesizing practical feedback control schemes for distributed parameter systems via functional analysis. Also needed is a sensitivity analysis of the optimal control algorithms reported here with respect to model inaccuracies and system's noise.

Further work is also needed to evaluate trade-offs in distributed model sophistication versus worth and practical implementability of the optimizing algorithms.

Index

Index